# Modeling and Optimization in Science and Technologies

Volume 20

The book series Modeling and Optimization in Science and Technologies (MOST) publishes basic principles as well as novel theories and methods in the fast-evolving field of modeling and optimization. Topics of interest include, but are not limited to: methods for analysis, design and control of complex systems, networks and machines; methods for analysis, visualization and management of large data sets; use of supercomputers for modeling complex systems; digital signal processing; molecular modeling; and tools and software solutions for different scientific and technological purposes. Special emphasis is given to publications discussing novel theories and practical solutions that, by overcoming the limitations of traditional methods, may successfully address modern scientific challenges, thus promoting scientific and technological progress. The series publishes monographs, contributed volumes and conference proceedings, as well as advanced textbooks. The main targets of the series are graduate students, researchers and professionals working at the forefront of their fields.

**Indexed by SCOPUS. All books published in the series are submitted for consideration in Web of Science.**

Samir Brahim Belhaouari · Halima Bensmail ·
Farshid Mehrdoust

# Discrete Mathematics, Probability Theory and Stochastic Processes

## For Applications in Engineering and Computer Science

Springer

Samir Brahim Belhaouari
College of Science and Engineering
Hamad Bin Khalifa University
Doha, Qatar

Halima Bensmail
Qatar Computing Research Institute
Hamad Bin Khalifa University
Doha, Qatar

Farshid Mehrdoust
Department of Applied Mathematics
University of Guilan
Rasht, Iran

ISSN 2196-7326 ISSN 2196-7334 (electronic)
Modeling and Optimization in Science and Technologies
ISBN 978-3-031-80588-2 ISBN 978-3-031-80589-9 (eBook)
https://doi.org/10.1007/978-3-031-80589-9



This Springer imprint is published by the registered company Springer Nature Switzerland AG
The registered company address is: Gewerbestrasse 11, 6330 Cham, Switzerland

# Preface

This book is designed to be a comprehensive and practical guide for computer scientists and engineers seeking to understand the fundamental concepts and applications of discrete mathematics and probability theory to problems relating to stochastic processes across a wide range of fields, including artificial intelligence (AI) and engineering.

Discrete mathematics, probability theory, and stochastic processes form the backbone of numerous applications in modern engineering, including AI. AI algorithms heavily rely on discrete structures and mathematical foundations to process and analyze data, make predictions, and learn from patterns. Thus, understanding discrete mathematics will provide engineers with the necessary background to design efficient algorithms, develop intelligent systems, and optimize decision-making processes.

Probability theory is indispensable in the field of AI, as it enables engineers to model and reason about uncertain events and outcomes. From machine learning algorithms that estimate probabilities to probabilistic graphical models that capture complex relationships, probability theory underpins many AI applications. Engineers equipped with a solid understanding of probability theory can make informed decisions, assess risk, and develop robust AI systems.

Stochastic processes play a crucial role in AI and engineering, as they capture the dynamics of systems that evolve over time [1–3]. In AI, stochastic processes are employed to model time-dependent phenomena, such as natural language processing, speech recognition, and financial forecasting. Engineers leverage stochastic processes to analyze system behavior, optimize resource allocation, and simulate real-world scenarios to improve AI algorithms and system performance.

Beyond AI, the concepts covered in this book have broad applicability in various engineering domains. Discrete mathematics finds applications in network design, cryptography, optimization, and information theory, enabling engineers to solve complex problems efficiently [4–6]. Probability theory is essential in fields such as telecommunications, reliability engineering, and signal processing [7–9], where uncertainty and random phenomena are prevalent. Stochastic processes find applications in areas such as control systems, operations research, and queuing theory, providing engineers with valuable tools to analyze and optimize system performance [10–12].

If you are looking for textbooks on probability and statistics that are appropriate for undergraduate students, you will find a variety of options on the market. These textbooks may come in various editions, including both old and new prints, but they all cover fundamental concepts. In particular, they are designed to provide a solid foundation in probability and statistics for students who are just beginning to explore these subjects. For those pursuing graduate studies in probability and statistics, there are specialized textbooks available that cover advanced topics and recent developments in the field.

Throughout this book, we will explore the practical applications of discrete mathematics, probability theory, and stochastic processes in AI, engineering, and related fields. We will dive into examples and case studies that highlight how these concepts are applied to tackle real-world challenges. By the end of this journey, you will have a solid foundation in these areas and be ready to apply them to a wide range of applications in AI, engineering, and beyond.

Doha, Qatar — Samir Brahim Belhaouari
Doha, Qatar — Halima Bensmail
Rasht, Iran — Farshid Mehrdoust
October 2024

## References

1. Scott M (2013) Applied stochastic processes in science and engineering, University of Waterloo, eBook
2. Gallager RG (2013) Stochastic processes: theory for applications. Cambridge University Press
3. Wong E, Hajek B (2012) Stochastic processes in engineering systems. Springer Science & Business Media
4. Rosen KH (2007) Discrete mathematics and its applications. The McGraw Hill Companies
5. Graham RL, Knuth DE, Patashnik O (1994) Concrete math: a foundation for computer science
6. Tucker A (1994) Applied combinatorics. John Wiley & Sons, Inc.
7. Peebles Jr PZ (1987) Probability Random variables, and random signal principles
8. Gubner JA (2006) Probability and random processes for electrical and computer engineers. Cambridge University Press
9. Kobayashi H, Mark BL, Turin W (2011) Probability, random processes, and statistical analysis: applications to communications, signal processing, queueing theory and mathematical finance. Cambridge University Press
10. Wong E, Hajek B (2012) Stochastic processes in engineering systems. Springer Science & Business Media
11. Jones PW, Smith P (2017) Stochastic processes: an introduction. CRC Press
12. Stirzaker D (2005) Stochastic processes and models. OUP Oxford

# Contents

# Chapter 1
# Topics in Mathematics and Applications

The course is not meant to be about the philosophy of mathematics, but it is important to realize that this question is still far from being understood. There are people whose only job is to investigate and discuss this topic. One should also understand that there cannot be one absolute all-encompassing answer: mathematics means different things to different people (even to different mathematicians). Here is a (non-exhaustive) sample of possible answers:

- Mathematics is the language of Science.
- Mathematics is the study of patterns.
- Mathematics is the study of topics such as quantity (numbers), structure, space, and change.
- Mathematics is the study of what can precisely be argued to be true or false.
- Mathematics is what mathematicians do.
- Mathematics is the subject in which we never know what we are talking about, nor whether what we are saying is true.

Whatever mathematics is, most mathematicians would agree that it involves explaining rather than just describing. The most highly-valued form of explanation in mathematics is called a proof.

In mathematics, a proof is a very precise argument explaining why a given statement is true.

The statement that is being proven, as well as every step of the proof, must be unambiguous: if there is ambiguity on what the statement even says, how can one agree about its truth? In particular, the proof should be understandable to its audience.

The proof must be logically sound: not only must every step be correct, but steps should also be justified so that no doubt is left about their validity.

A proof must rely on common ground shared by the entire audience: if the audience disagrees on the truth of every single fact, including whether $1+1=2$ or $1=1$, then there is no hope of convincing it using pure reason. This common ground includes some (hopefully simple) statement whose truth is taken as granted (the axioms).



S. Brahim Belhaouari et al., *Discrete Mathematics, Probability Theory and Stochastic Processes*, Modeling and Optimization in Science and Technologies 20,
https://doi.org/10.1007/978-3-031-80589-9_1

Now look at a very different style of argument. We say a number $x$ is non-negative if $x \geq 0$.

**Theorem** *For all non-negative real numbers a and b* : $a^2 + b^2 \leq (a + b)^2$

***Proof 1***

- $a^2 + b^2 \leq (a + b)^2$
- $a^2 + b^2 \leq a^2 + b^2 + 2ab$
- $0 \leq 2ab$
- True, because the product of two non-negative numbers is non-negative.

***Proof 2***

- Since the product of non-negative numbers is non-negative, $0 \leq 2ab$,
- adding $a^2 + b^2$ to both sides, one gets: $a^2 + b^2 \leq a^2 + b^2 + 2ab$.
- So factoring the right hand side: $a^2 + b^2 \leq (a + b)^2$,
- which is the desired inequality.

## 1.1 Methods of Proofs

Most of the theorems we want to prove are as follows:

$$P \Rightarrow Q$$

where Q and P are specified statements.(Fig. 1.1).

### *1.1.1 Direct Proofs*

Direct proof is a method of proof that starts with $P$ and ends with $Q$.

***Example*** The sum of two even integers equals an even integer.

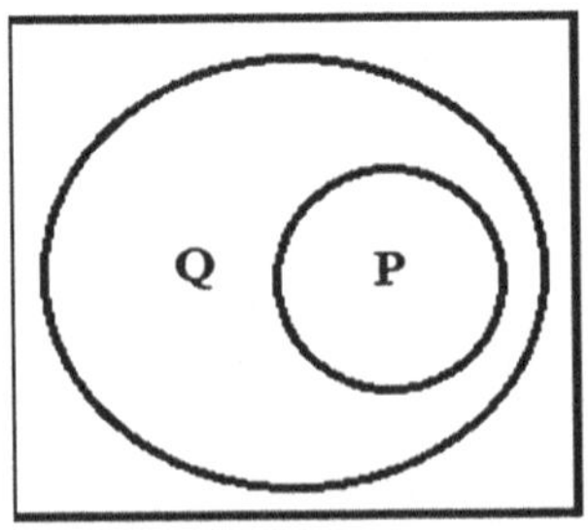


**Fig. 1.1** The statements $Q$ and $P$ to prove $P \Rightarrow Q$

***Proof*** Consider two even integers $x$ and $y$. Since they are even, they can be written as

$$x = 2a, y = 2b,$$

respectively, for integers $a$ and $b$. Then the sum can be written as

$$x + y = 2a + 2b = 2(a + b) = 2p,$$

where, $p = a + b$, $a$ and $b$ are all integers.

It follows that $x + y$ has 2 as a factor and therefore is even, so the sum of any two even integers is even.

### *1.1.2 Proof by Contradiction*

Proof by contradiction is a method of mathematical proof where one assumes the negation of what they are trying to prove and then derives a contradiction. This contradiction proves that the assumption of the negation is false, so the original statement must be true. Here is an example of a proof by contradiction: Statement to prove: $\sqrt{2}$ is irrational Assume the opposite: $\sqrt{2}$ is rational If $\sqrt{2}$ is rational, then it can be expressed as a ratio of two integers: $\sqrt{2} = a/b$ where a and b are integers with no common factors. Squaring both sides, we get $2 = a^2/b^2$ This implies that $2b^2 = a^2$, so $a^2$ must be even. Since the square of an odd number is always odd, this means that a must be even. So we can write a = 2n where $n$ is an integer. Substituting back into the original equation, we get $2b^2 = (2n)^2 = 4n^2$. This simplifies to $b^2 = 2n^2$, so $b^2$ must also be even. Therefore, both $a$ and $b$ are even, which means they have a common factor of 2. This contradicts our initial assumption that $a$ and $b$ have no common factors. Since assuming the opposite led to a contradiction, we can conclude that $\sqrt{2}$ is indeed irrational. This is a proof by contradiction.

In this method of proof, we consider the result we need to prove, that is, the expression $Q$, to be incorrect.

Thus, we assume $P$ and not $Q$ (denote as ~ $Q$). In fact, in this method of proof, we must conclude that it either contradicts with problem assumptions or is obviously not true.

$P \wedge \sim Q =$ False. (Fig. 1.2)

**Theorem** *If x is integer and $x^2$ is even, then x is even, i.e.,*

$$\forall x \in \mathbb{Z} \wedge x^2 \text{ is even } \Rightarrow x \text{ is even.}$$

***Proof***

$$\forall x \in \mathbb{Z} \wedge x^2 \text{is even } \wedge x \text{ is odd } \Rightarrow \text{False.}$$

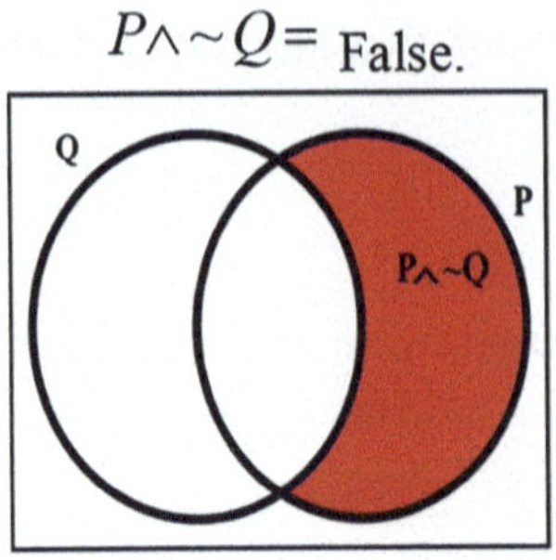


**Fig. 1.2** The statements $Q$ and $P$ to prove $P\wedge \sim Q =$ False

Since, $x^2$ is an even integer, $\exists k \in \mathbb{Z} : x^2 = 2k$. On the other hand, since $x$ is an even integer, $\exists n \in \mathbb{Z} : x = 2n + 1$. Combine them together

$$\begin{aligned} 2k &= (2n+1)^2 \\ 2k &= 4n^2 + 4n + 1 \\ 2\left(k - 2n^2 - 2n\right) &= 1\cdot \end{aligned}$$

In the contradiction method of proof we consider the opposite of the logical concept

$$P\wedge \sim Q = \text{False}.$$

In the proof by contradiction method, the purpose of the proof is clear: we must prove the expression $\sim P$.

**Theorem** *If x and y are two integers whose product is odd, then both must be odd, i.e.,*

$$x, y \in \mathbb{Z} \wedge x \times y \,\text{is odd} \Rightarrow x \wedge y \,\text{are odd}.$$

***Proof***

$$\text{either}\, x \vee y \,\text{is even} \Rightarrow x \times y \,\text{is even}.$$

Assume that $x = 2k$ or $y = 2n$. Then

$$\begin{aligned} &x = 2k, y = 2n + 1 \Rightarrow xy = 2k(2n+1), \\ &x = 2k, y = 2n \Rightarrow xy = 4nk, \\ &x = 2k + 1, y = 2n \Rightarrow xy = 2n(2k+1). \end{aligned}$$

### 1.1.3 Proof by Mathematical Induction

Proof by mathematical induction is a method used to prove that a statement is true for all positive integers. The steps involved in a proof by mathematical induction are as follows: Base Case: The statement is first proved to be true for the smallest possible value of the variable in question, typically $n = 1$ or $n = 0$. Inductive Hypothesis: Assume that the statement is true for some arbitrary positive integer $k$. This is known as the inductive hypothesis. Inductive Step: Using the inductive hypothesis, the statement is proved to be true for the next value of the variable, which is $k + 1$. By following these steps, the proof establishes that the statement is true for all positive integers. Mathematical induction is a powerful tool in mathematics and is frequently used to prove statements pertaining to sequences, series, inequalities, and divisibility.

This proof technique can be stated as

$$[P(n_0) \wedge \forall k(P(k) \Rightarrow P(k+1))] \Rightarrow \forall n \geq n_0 P(n).$$

We first prove that the theorem is true of the first case. Then we prove that if it is true for one case, it is true for the next. This proves to be true for all positive integers.

***Example*** Prove by induction that $11^n - 6$ is divisible by 5 for every positive integer n.

*Solution* Let $P(n)$ be the mathematical statement

$$11^n - 6 \text{ is divisible by } 5\cdot$$

Base Case: When $n = 1$, we have $11^1 - 6 = 5$ which is divisible by 5. So $P(1)$ is correct.

Induction hypothesis: Assume that $P(k)$ is correct for some positive integer $k$. That means $11^k - 6$ is divisible by 5 and, hence, $11^n - 6 = 5m$ for some integer m. So $11^k = 5m + 6$.

Induction step: We will now show that $P(k+1)$ is correct. Always keep in mind what we are aiming for and what we know to be true. In this case we want to show that $11^{k+1} - 6$ can be expressed as a multiple of 5, so we will start with the formula and we will rearrange it into something involving multiples of 5. At some point we will also want to use the assumption that $11^k = 5m + 6$.

$$\begin{aligned}
11^{k+1} - 6 &= \left(11 \times 11^k\right) - 6 && \text{by the laws of powers} \\
&= 11(5m + 6) - 6 && \text{by the inductuin hypothesis} \\
&= 11(5m) + 66 - 6 && \text{by expanding the bracket} \\
&= 5(11m) + 60 \\
&= 5(11m + 12) && \text{since both parts of the formulahave a common factor of 5.}
\end{aligned}$$

As $11m + 12$ is an integer we have that $11^{k+1} - 6$ is divisible by 5, so $P(k+1)$ is correct. Hence, by mathematical induction $P(n)$ is correct for all positive integers n.

***Example*** Prove by induction that $2^n > 2n$ for every positive integer $n > 2$.

*Solution* Let $P(n)$ be the mathematical statement.

Base Case: When $n = 3$ we have $2^3 = 8 > 6 = 2 \times 3$. So $P(3)$ is correct. Induction hypothesis: Assume that $P(k)$ is correct for some positive integer $k$. That means that $2^k > 2k$.

Induction step: We will now show that $P(k+1)$ is correct.

$$\begin{aligned} 2^{k+1} &= 2 \times 2^k > 2 \times 2k \quad \textit{by the induction hypothesis} \\ &= 2(k+1). \end{aligned}$$

So $P(k+1)$ is correct. Hence, by mathematical induction $P(n)$ is correct for all positive integers $n > 2$.

## 1.2 Sequences and Series

We begin with a sequence of even numbers as our first mathematical experiment

$$2, 4, 6, 8, 10, ...$$

and consider odd numbers

$$1, 3, 5, 7, 9, ...$$

In this case, we can use common sense to "predict" the twentieth part of each sequence.

Another sequel that has received a lot of attention over the years is the Fibonacci sequence

$$1, 1, 2, 3, 5, 8, 13, 21, 34, 55, 89, 144, ...$$

where each expression is the sum of the previous two expressions. For instance, $55 = 21 + 34$. In this case, it is much more difficult to predict the twentieth period, without mentioning all the previous periods.

Sequences are used in many areas of mathematics, including finance. For example, we can invest \$1000 a year at an annual interest rate of 5% and list the annual value of the investment as follows:

$$\$1000, \$1050, \$1102.50, \$1157.63, \$1215.51, ...$$

Sequences can be finite or infinite. For instance,

$$2, 4, 6, 8, 10$$

is a finite sequence with five expressions, while

$$2, 4, 6, 8, 10, \ldots$$

continues without limitation and is an infinite sequence. We usually use "..." to indicate that the sequence continues indefinitely.

For a specified sequence, we can propose the following statements:

- Is there a general formula for finding a sequence?
- Do the sequences approach an arbitrary number and have a limit?

For example, we can see intuitively that the terms in the infinite sequence

$$1, \frac{1}{2}, \frac{1}{3}, \frac{1}{4}, \frac{1}{5}, \ldots$$

whose general term is $\frac{1}{n}$, are approaching 0 as $n$ becomes very large.

A finite series arises when we add the terms of a finite sequence. For example,

$$2 + 4 + 6 + 8 + \cdots + 20$$

is the series formed from the sequence 2, 4, 6, 8, ..., 20.

An infinite series is the 'formal sum' of the terms of an infinite sequence. For example,

$$1 + 3 + 5 + 7 + 9 + \ldots$$

is the series formed from the sequence of odd numbers. We can spot an interesting pattern in this series. The sum of the first two terms is 4, the sum of the first three terms is 9, and the sum of the first four terms is 16. So we guess that, in general, the sum of the first $n$ terms is $n^2$.

For a specified infinite sequence, we can propose the following questions:

- Can we find a formula for the sum of the $n$ first terms of the series?
- Does this series have a limit, that is, if we add $n$ terms to the first expression of the series, does this value get closer to any number as $n$ gets bigger?

If present, this limit is often referred to as the limit of the infinite series. In this module, we examine the limiting amounts for a specific but commonly known series, known as the geometric series.

### 1.2.1 Sequences

The list of positive odd numbers

$$1, 3, 5, 7, 9, ...$$

is an example of a typical infinite sequence. The 3 dots indicate that the sequence continues forever, with no last term. We will use the symbol $a_n$ to denote the $n^{th}$ term of a given sequence. Thus, in this example, $a_1 = 1, a_2 = 3, a_3 = 5$ and so on; the first term is $a_1 = 1$, but there is no last term.

The list of positive odd numbers less than 100

$$1, 3, 5, 7, ..., 99$$

is an example of a typical finite sequence. The first term of this sequence is 1 and the last term is 99. This sequence contains 50 terms.

There are several ways to display a sequence:

- Write out the first few terms.
- Give a formula for the general term.
- Give a recurrence relation.

Writing the first few terms is not a good idea, because you have to "believe" that there is a certain pattern and that there may be many such patterns. For instance, if we simply write

$$1, 2, 4, ...$$

If there is a more complex pattern, the next expression may be 8 or possibly 7 or maybe 23. Therefore, if only the first few terms are mentioned, there should be some rules about how to uniquely determine the next term in the sequence.

A much better way to describe a sequence is to give a formula for the $n$th term $a_n$. This is also called a formula for the general term. For example, a better approach to describing a sequence is to have a formula available for the $n$th expression $a_n$. This formula is also called a general term. For instance,

$$a_n = 2n - 1, \; for \; n \in \mathbb{N} - \{0\}.$$

is a formula for the general term in the sequence of odd numbers $1, 3, 5, ....$ From the formula, we can, for example, write down the 10th term, since $a_{10} = 2 \times 10^{-1} = 19$.

In some cases, providing an explicit formula for $a_n$ is not easy or even possible. In such cases, it may be possible to define a particular term in order of some of the previous terms. This expression is frequently referred to as a recurrence relation. For example, the sequence of odd numbers may be

$a_1 = 1$ and $a_{n+1} = a_n + 2$, for $n \geq 1$.

$a_1 = 1$ is the initial term, and from recurrence relationship we must add two numbers to each expression to get the next expression.

As mentioned, the Fibonacci sequence consists of the following numbers

$$1, 1, 2, 3, 5, 8, 13, 21, 34,...$$

where any expression is the sum of the previous two expressions. This can be expressed by setting $a_1 = a_2 = 1$ and $a_{n+2} = a_{n+1} + a_n$, for $n \geq 1$.

In general, determining a formula for the general expression of a sequence can be difficult. Take the Fibonacci sequence, for example:

$$1, 1, 2, 3, 5, 8, 13, 21, 34, ....$$

In the following sections, we will discuss how to obtain the general term for the Fibonacci sequence as follows:

$$a_n = \frac{1}{\sqrt{5}}\left(\left(\frac{1+\sqrt{5}}{2}\right)^n - \left(\frac{1-\sqrt{5}}{2}\right)^n\right).$$

This is a very amazing result! (It is not even clear that this formula will give a correct result for every $n$). You may want to check whether this formula gives the desired result for each $n = 1, 2, 3$.

In addition, we can use sequences to approximate real numbers. For example, the following sequence gives an approximation of the real number $\sqrt{2}$.

$$1, 1.4, 1.41, 1.414, 1.4142, ...$$

### 1.2.2 Arithmetic Sequences

We will limit our attention for the moment to one particular type of sequence, known as an arithmetic sequence (or arithmetic progression). This is a sequence of the form

$$a, a + d, a + 3d, ...$$

where each term, including the previous one, is obtained by adding a constant, called a common difference, and it is often denoted by a symbol $d$. Note that $d$ can be negative, or zero (Fig. 1.3).

Therefore, the sequence of even numbers

$$2, 4, 6, 8, 10, ...$$

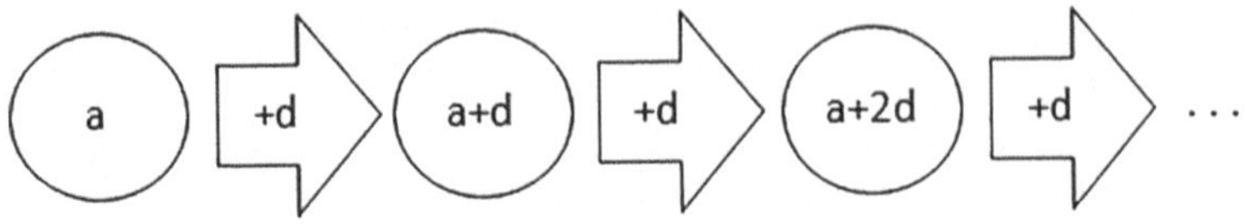


**Fig. 1.3** A diagram for an arithmetic sequence

is an arithmetic sequence whose common difference is $d = 2$.

It can be easily inferred that the formula for the $n$th sentence of arithmetic sequences is expressed as follows:

$$a_n = a_0 + (n-1)d, n \geq 1.$$

***Example*** Derive the formula for the $n$th term of the arithmetic sequence.

1. $2, 5, 8, \ldots$
2. $107, 98, 89, \ldots$

*Solution*

1. Here $a_0 = 2$ and $d = 3$, so

   $a_n = 2 + (n-1) \times 3 = 3n - 1$

2. Here $a_0 = 107$ and $d = -9$, so

   $a_n = 107 + (n-1) \times (-9) = 116 - 9n$

Moreover, we can examine whether a specified number belongs to a certain arithmetic sequence.

***Example*** Does the number 203 belong to the arithmetic sequence 3, 7, 11, ...?

*Solution* Here $a = 3$ and $d = 4$, so $a_n = 3 + (n-1) \times 4 = 4n - 1$. According to $4n - 1 = 203$, we find $n = 51$. Thus, 203 is the $51st$ expression of the sequence.

### *1.2.3 Geometric Sequence*

The general form of a geometric sequence is as follows:

$$a, ar, ar^2, ar^3, \ldots$$

where each sentence is obtained by multiplying by a constant, called the common ratio, often denoted by the symbol $r$, from the previous term. Note that r can be

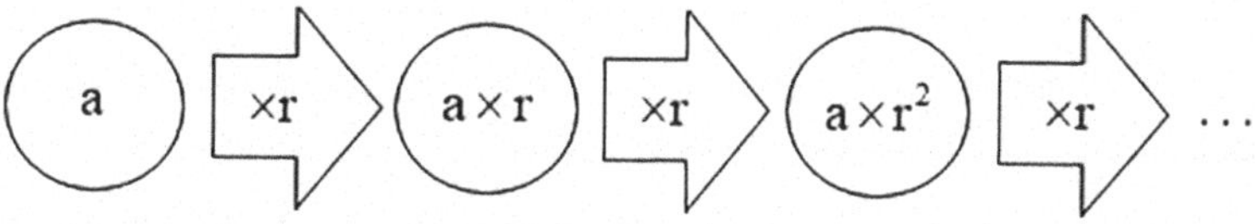


**Fig. 1.4** A diagram for a geometric sequence

negative, positive or zero. Further, in a geometric sequence with negative common ratio, the sentences of sequence fluctuate between negative and positive (Fig. 1.4).

The following sequence

$$1, 2, 4, 8, 16, 32, 64, ...$$

is an example of a geometric sequence with initial term 1 and common ratio $r = 2$, while

$$3, -6, 12, -24, 48, -96, ...$$

is an sample of a geometric sequence with initial term 3 and common ratio $r = -2$. It can be easily inferred that the formula for the $n$th sentence of geometric sequences is expressed as follows:

$$a_n = a_1 r^{n-1}, \forall n \geq 1; a_n = a_0 r^n, \forall n \geq 0\cdot$$

***Example*** Derive the formula for the $n^{th}$ sentence of the following geometric sequence.

1. $2, 6, 18, ...$
2. $486, 162, 54, ...$

*Solution*

1. Here $a_1 = 2$ and $r = 3$, so $a_n = 2 \times 3^{n-1}$.
2. Here $a_1 = 486$ and $r = \frac{1}{3}$, so $a_n = 486 \times \left(\frac{1}{3}\right)^{n-1}$.

***Example*** Does the number 48 belong to the geometric sequence
3072, 1536, 768, ...?

*Solution* Here $a_1 = 3072$ and $r = \frac{1}{2}$ so $a_n = 3072 \times \left(\frac{1}{2}\right)^{n-1} = \frac{1}{64}$ that is, $2^{n-1} = 64 = 2^6$, and so $n = 7$. Hence, 48 is the 7th term of the sequence.

***Example*** Does the number 6072 belong to the following geometric sequence

$$3, -6, 12, -24, 48, ...?$$

*Solution*

Here $a_1 = 3$ and $r = -2$, so $a_n = 3 \times (-2)^{n-1}$. We set $3 \times (-2)^{n-1} = 6072$. This gives $(-2)^{n-1} = 2024$. But 2024 is not a power of 2, and so 6072 does not belong to the sequence.

### *1.2.4 Series*

A finite series is the summand of the sentences of a finite sequence. Therefore, if

$$a_1, a_2, \ldots, a_n$$

is a sequence of $n$ sentences, then the corresponding series is

$$a_1 + a_2 + \cdots + a_n$$

The number $a_k$ is referred to as the $k$th term of the series.

We mostly use the Sigma symbol for series. For instance, if we have a set in which the $k$th expression is given by $2k$, we can write this series as follows:

$$\sum_{k=1}^{50} 2k$$

We note that the variable $k$ here is a fictitious variable. This means that the above series can written as follows:

$$\sum_{i=1}^{50} 2i \text{ or } \sum_{j=1}^{50} 2j.$$

An infinite series can be expressed as the sum of an infinite sequence of the following form

$$a_1 + a_2 + a_4 + \ldots$$

For instance, the odd numbers sequence leads the infinite series as $1+3+5+7+\ldots$. We can also write an infinite series of finite sentences. In other words, the sum of the first n sentences of an infinite series can be written as follows:

$$S_n = a_1 + a_2 + \cdots + a_n = \sum_{k=1}^{n} a_k$$

This expression is sometimes called the $n$th partial sum of the infinite series. Based on the formula for the sum of the first $n$ expressions of a specified series, we can retrieve a formula for this expression with a simple subtraction as follows:

$$S_n = a_1 + a_2 + \cdots + a_{n-1} + a_n,$$

$$S_{n-1} = a_1 + a_2 + \cdots + a_{n-1},$$

by subtracting we obtain

$$S_n - S_{n-1} = a_n.$$

For instance, if the partial sum of the first $n$ sentences for a specified series is as $S_n = n^2$, then the $n$th sentence can be expressed as follows:

$$a_n = S_n - S_{n-1} = n^2 - (n-1)^2 = 2n - 1\cdot$$

In the other words, the sentences in this sequence are as odd numbers. Therefore, we have found a formula for the sum of the first $n$ integer numbers:

$$1 + 3 + 5 + \cdots + (2n-1) = n^2$$

In general, it is difficult to obtain a closed form for the sum of a series with $n$ sentences. In the following of this section, the arithmetic and geometric series are considered.

### *1.2.5 Arithmetic Series*

A series whose sentences form an arithmetic sequence is called an arithmetic series. In other words, the sequence in an arithmetic series is obtained by adding a constant from the previous sentence.

The series

$$1 + 2 + 3 + \cdots + n$$

represents an arithmetic series with common difference 1. In order to find the sum of this series, we consider the series forwards and next backwards as follows:

$$S_n = 1 + 2 + 3 + \cdots + (n-1) + n,$$
$$S_n = n + (n-1) + (n-2) + \cdots + 2 + 1.$$

Adding both sides, we have

$$2S_n = (1+n) + (2+n-1) + (3+n-2) + \cdots + (n-1+2) + (n+1)\cdot$$

The expression on the right hand side equals to $n(n+1)$. Therefore, $2S_n = n(n+1)$, that is,

$$1 + 2 + 3 + \cdots + n = \frac{1}{2}n(n+1).$$

For instance,

$$1 + 2 + 3 + \cdots + 100 = \frac{1}{2} \times 100 \times 101 = 5050.$$

This formula works for any arithmetic series and provides the sum of the first $n$ sentence of an arithmetic series based on first sentence $a_1 = a$ and last sentence $a_n = \ell$. Thus, we have.

$$S_n = \frac{n}{2}(a + \ell)$$

Since the last term can be written as $a_n = a_1 + (n-1)d$, where $d$ is the common difference, we also have

$$\begin{aligned} S_n &= \frac{n}{2}(a + \ell) \\ &= \frac{n}{2}(a + a + (n-1)d) \\ &= \frac{n}{2}(2a + (n-1)d). \end{aligned}$$

***Example*** Derive the general formula for the sum of the first $n$ sentences of the following arithmetic sequence

1. 2, 5, 8, ...
2. 107, 98, 89, ...

*Solution*

1. Here $a_1 = 2$ and $d = 3$, so

$$S_n = \frac{n}{2}(4 + (n-1) \times 3) = \frac{n}{2}(3n+1).$$

We can gain the $n$th sentence of the series, which is $a_n = 3n - 1$, and apply the following formula

$$S_n = \frac{n}{2}(a_1 + \ell) = \frac{n}{2}(2 + (3n-1)) = \frac{n}{2}(3n+1).$$

2. Here $a_1 = 107$ and $d = -9$, so

$$S_n = \frac{n}{2}(2 \times 107 + (n-1) \times (-9)) = \frac{n}{2}(223 - 9n).$$

In the previous example, we can put $n = 1$ and check this is the first sentence of the series. Note that since the sum formula is a a quadratic, examining three cases $n = 1, n = 2, n = 3$ is sufficient to prove the correct answer.

### 1.2.6 Geometric Series

A series whose sentences form a geometric sequence is called a geometric series, i.e., each sentence is multiplied by a constant.

For instance,

$$2 + 8 + 32 + 128 + \ldots$$

represents a geometric series with common ratio 4 and first sentence 2. Further, the $n$th term is $a_n = 2 \times 4^{n-1}$.

Based on a small trick, we can find a formula for the sum of the first $n$ sentences of this series. For this purpose, multiply the series by the common ratio of 4 and then subtract the original equation as follows:

$$S_n = 2 + 8 + 32 + 128 + \cdots + 2 \times 4^{n-1}$$
$$4S_n = 8 + 32 + 128 + \cdots + 2 \times 4^{n-1} + 2 \times 4^n,$$

We obtain

$$4S_n - S_n = 2 \times 4^n - 2$$

And

$$S_n = \frac{1}{3}\left(2 \times 4^n - 2\right) = \frac{2(4^n - 1)}{3}.$$

This method can be used for any geometric series and the general formula for the sum of the first $n$ sentences of a geometric series with the first sentence $a$ and the common ratio $r$. So we have the following formula

$$S_n = \frac{a(r^n - 1)}{r - 1} for \quad r \neq 1.$$

This can also be written as follows:

$$S_n = \frac{a(1-r^n)}{1-r} for \quad r \neq 1.$$

When r is between −1 and 1, it is often easier to use the second formula.

The second formula is often more convenient to use when r lies between −1 and 1. Moreover, for $r = 1$, the sum of the series is clearly $na$, because all terms are same.

***Example*** Derive a formula for the summand of the first $n$ sentences of the following geometric sequences.

1. 2, 6, 18, ...,
2. 486, 162, 54, ....

*Solution*

1. Here $a_1 = 2$ and $r = 3$, so

$$S_n = \frac{2(3^n-1)}{3-1} = 3^n - 1.$$

2. Here $a_1 = 486$ and $r = \frac{1}{3}$, so

$$S_n = \frac{486\left(1-\left(\frac{1}{3}\right)^n\right)}{1-\frac{1}{3}} = 729\left(1-\left(\frac{1}{3}\right)^n\right).$$

In the previous example, we can substitute $n = 1$ and examine that we gain the first sentence of the sequence.

### 1.2.7 Means

Finding the mean of two numbers is a simple matter. For instance, consider the mean of 6 and 10 is 8. When we do this, we actually find an x that $6, x, 10$ forms an arithmetic sequence. In general, if $a, x, b$ is an arithmetic sequence, then

$$\frac{x}{a} = \frac{b}{x} \cdots \Rightarrow \cdots x = \frac{a+b}{2}.$$

This is the mean of $a$ and $b$, which is also called the arithmetic mean $a$ and $b$. Similarly, we can define the geometric mean of two positive numbers $a$ and $b$, the positive number $x$, and $a$, $x$, $b$ such that it forms a geometric sequence. For this purpose, we need

$$\frac{x}{a} = \frac{b}{x} \cdots \Rightarrow \cdots x = \sqrt{ab}.$$

***Example*** Find the arithmetic and geometric means of the following numbers

1. 2, 18
2. 3, 6.

*Solution*

1. $AM = \frac{2+18}{2} = 10, GM = \sqrt{2 \times 18} = 6.$
2. $AM = \frac{3+6}{2} = \frac{9}{2}, GM = \sqrt{3 \times 6} = 3\sqrt{2}.$

## *1.2.8 Applications to Finance*

In problems related to financial mathematics, we can see the applications of the use of sequences and series in abundance. In the following, we study compound interest and superannuation.

### 1.2.8.1 Compound Interest

Compound interest is a type of interest in which the amount of interest is added to the principal of capital and in subsequent rounds, in addition to the principle of investment, interest is also accrued on interest. The greater the number of combined courses, the greater the compound interest. Therefore, the amount of compound interest that is given to a loan of one million dollars annually at a rate of 10% is less than the amount of compound interest that is given to the same loan of one million dollars, but this time at a rate of 5% and on a six-month basis. Although the time interval of both is a total of one year.

Assume that the time interval is 1 year. Thus, after 1 year, the investor has a capital $P + rP = P(1 + r)$ with interest rate $r$. In the same way, after 2 years, the investor has a capital $P(1 + r) + rP(1 + r)^2$. As a result, it can be seen that after $n$ years, the total investment will be $P(1 + r)^n$.

These amounts form a geometric sequence with common ratio $1 + r$, and the $n$th term as follows:

$$P(1 + r)^n.$$

***Example*** What is the value of capital after 10 years with an investment of \$1000 and an annual interest rate of 4%?

*Solution* We have $P = 2000$, $r = 0.04$ and $n = 10$. Therefore, the investment after 10 years, have amounts

$$2000(1 + 0.04)^{10} \approx \$2560.48.$$

Depreciation is closely related to compound interest. For example, a company can claim the depreciation of a car over time as a tax deduction.

Consider a car whose value is initially \$P so that it is depreciated annually at the rate of r, the value of this car after 1 year is $P - rP = P(1 - r)$. After 2 years, the value of the car is equal to $P(1 - r) - rp(1 - r) = P(1 - r)^2$. We can conclude that after $n$ years, the value of the car will be $P(1 - r)^n$

***Example*** Consider a car whose value is initially \$30,000 so that it is depreciated annually at the rate of 6%. Find the value of this car after 10 years.

Solution We have $P = 30000$, $r = 0.06$ and $n = 10$. Therefore, the worth of car after 10 years is

$$3000(1 - 0.06)^{10} \approx \$16158.$$

#### 1.2.8.2 Superannuation

Superannuation is a way to save for retirement. In fact, in the long run, money is invested regularly and its (compound) interest is paid. Consider a person who has invested in a retirement plan for 30 years and pays 6% per annum. In this case, he (she) would spend \$3000 a year on the plan, adding an annual profit. Thus, the first \$3000 will be invested for the full 30 years at 6% per annum. Therefore, it is awarded to $\$3000 \times 1.06^{30}$. The next \$3000, which is deposited at the beginning of the second year, is invested for 29 years and is awarded to $\$3000 \times 1.06^{30}$, and so on. Considering amounts from the smallest to the largest, the total value of his investment after 30 years is equal to

$$A = 3000 \times 1.06 + 3000 \times 1.06^2 + \cdots + 3000 \times 1.06^{30}.$$

This problem represents a geometric series with first sentence $a = 3000 \times 1.06$ and common ratio $r = 1.06$. Further the number of sentences is $n = 30$. Applying the formula for the summand of a geometric series with finite terms, the final value of the investment is expressed as follows:

$$A \approx \$251405.$$

This example shows both the value of regular savings and compound interest. Based on this problem, the \$90,000 invested over 25 years becomes approximately \$251,400.

### 1.2.9 The Limiting Sum of a Geometric Series

In mathematics, the concept of infinity and limits are closely related and fundamental to the understanding of calculus and other areas of mathematics. Infinity is a concept that represents something without bound or limit, or something that is infinitely large or infinitely small. In mathematics, infinity is often used to describe things that are unbounded or unlimited, such as the values of numbers that can approach infinity or zero. Limits, on the other hand, refer to the value that a function or sequence approaches as the input approaches a certain value. The concept of limits allows us to study the behavior of functions or sequences as they get closer to a certain point, without actually reaching that point. The concept of infinity and limits are often used together in calculus to describe functions and their behavior at points where they may not be defined or where they approach infinity. Understanding the concept of infinity and limits is crucial for understanding the behavior of functions and sequences in mathematics.

**Definition (limit).** Suppose that the function $f$ is defined in the neighborhood of $x_0$ except possibly $x_0$ itself. We say that the limit of the function $f$ that approaches $x_0$ at $x$ is equal to the real number $I$, whenever we want to approach the function $f$ to $I$ by any amount, provided that $x$ approaches $x_0$ sufficiently from both sides. In this case, we write: $\lim\limits_{x \to x_0} f(x) = I$. The value of $I$ is called the limit of function $f$ at $x_0$.

**Definition (limit).** Infinity is a concept in mathematics that represents a quantity or number that is larger than any finite number. It is often denoted by the symbol $\infty$.

One example of infinity in math is the limit of the sequence of natural numbers. As $n$ approaches infinity, the sequence $1, 2, 3, 4, \ldots$ continues to increase without bound. In this case, infinity represents the idea of numbers going on forever without reaching a maximum value.

**Definition (infinite series).** To introduce an infinite series, first consider a sequence as $\{a_n\}_{n=1}^{\infty}$. If the terms of this sequence are added as below, the infinite series sn will be formed as follows

$$\begin{aligned}
s_1 &= a_1 \\
s_2 &= a_1 + a_2 \\
s_3 &= a_1 + a_2 + a_3 \\
&\vdots \\
s_n &= a_1 + a_2 + a_3 + \cdots + a_n = \sum_{i=1}^{n} a_i
\end{aligned}$$

The term $s_n$ is called the partial sum. Therefore, the infinite series $s_n$ consists of the sum of terms of the sequence $a_n$. In this section, we want to specifically talk about infinite series. Note that an infinite series can also be represented as $\{s_n\}_{n=1}^{\infty}$. The result of the infinite series is obtained using the following limit

$$\lim_{n\to\infty} s_n = \lim_{n\to\infty} \sum_{i=1}^{n} a_i = \sum_{i=1}^{\infty} a_i.$$

The sum of an infinite series is the total value that the series converges to when all of its terms are added together. It is denoted by the symbol $\sum$ and can be written as:

$$\sum_{k=1}^{\infty} a_k = a_1 + a_2 + a_3 + \cdots$$

where $a_n$ represents the $n$th term in the series. The sum of an infinite series is said to converge if the partial sums of the series approach a fixed value as more terms are added. If the sum approaches infinity or does not converge to a fixed value, the series is said to diverge.

According to the previous section, sum of the first $n$ terms of a geometric series with common ratio r and first term a is

$$S_n = \frac{a(1-r^n)}{1-r} \text{ for } \; r \neq 1.$$

In the case when $r$ has magnitude less than 1, the term $r^n$ approaches 0 as $n$ becomes very large. So, in this case, the sequence of partial sums $S_1, S_2, S_3, \ldots$ has a limit:

$$\lim_{n\to\infty} S_n = \lim_{n\to\infty} \frac{a(1-r^n)}{1-r} = \frac{a}{1-r}.$$

The value obtained from the calculation of this limit is called the limiting sum of the infinite geometric series. As long as the $n$ is large enough, the partial sum $S_n$ of the series is as close as possible to the limiting sum.

The limiting sum is called the sum to infinity of a specified series and it is denoted by $S_\infty$. Thus, for a geometric series with common ratio $|r| < 1$,

$$S_\infty = \lim_{n\to\infty} S_n = \frac{a}{1-r}.$$

***Example*** Derive the limiting sum of the following geometric series

1. $1 + \frac{1}{3} + \frac{1}{9} + \cdots$
2. $8 - 6 + \frac{9}{2} - \cdots$

*Solution*

1. We have $a = 1$ and $r = \frac{1}{3}$, so the limiting sum exists, and it is as follows:

$$S_\infty = \frac{1}{1-\frac{1}{3}} = \frac{3}{2}.$$

2. We have $a = 8$ and $r = -\frac{3}{4}$, so the limiting sum exists, and

$$S_\infty = \frac{8}{1+\frac{3}{4}} = \frac{32}{7}.$$

We need to calculate the limit of the sum $a_1 + a_2 + a_3 + ....$ This is achieved by calculating the limit on the sequence of partial sums $S_1, S_2, S_3....$, where

$$\begin{aligned} S_1 &= a_1 \\ S_2 &= a_1 + a_2 \\ S_3 &= a_1 + a_2 + a_3 \end{aligned}$$

and so on. Infinite series are often expressed as follows:

$$\sum_{n=1}^{\infty} a_n$$

If the series has a limiting sum L, then the infinite series converges to L. This is expressed as

$$\sum_{n=1}^{\infty} a_n = L \; or \; a_1 + a_2 + a_3 + ... = L.$$

## 1.3 Links Forward

### *1.3.1 Use of Indication*

As we found out, it is easy to sum of certain series by recognizing patterns. For example, we saw that the sum of the first $n$ odd numbers is obtained as follows:

$$1 + 3 + 5 + \cdots + (2n - 1) = n^2.$$

On the other hand, the formula for the sum of the first $n$ squares,

$$1^2 + 2^2 + 3^2 + \cdots + n^2 = \frac{1}{6}n(n+1)(2n+1),$$

is hardly obvious and requires a proof. While the Greeks had some very creative proofs of results such as this, the best approach is to give a proof using mathematical induction.

#### 1.3.1.1 Telescoping Series

Most series are neither geometric nor arithmetic. Some of these series can be expressed by adding the difference of terms.

***Example***

1. Derive the sum

$$\sum_{k=2}^{n} \frac{2}{k^2 - 1}.$$

2. Does the following infinite series have a limiting sum?

$$\sum_{k=2}^{n} \frac{2}{k^2 - 1}$$

*Solution* We have

$$\begin{aligned}\frac{2}{k^2 - 1} &= \frac{2}{(k-1)(k+1)} = \frac{A}{k-1} + \frac{B}{k+1}\\ &= \frac{A(k+1) + B(k-1)}{(k-1)(k+1)}.\end{aligned}$$

On the other hand, $A(k+1) + B(k-1) = 2$. So $A = 1$ *and* $B = -1$. Thus,

$$\frac{2}{k^2 - 1} = \frac{1}{k-1} - \frac{1}{k+1}.$$

Therefore,

$$\begin{aligned}\sum_{k=2}^{n} \frac{2}{k^2 - 1} &= \sum_{k=2}^{n} \left( \frac{1}{k-1} - \frac{1}{k+1} \right)\\ &= \sum_{k=2}^{n} \frac{1}{k-1} - \sum_{k=2}^{n} \frac{1}{k+1}.\end{aligned}$$

Then we can write

$$\left(1 + \frac{1}{2} + \frac{1}{3} + \frac{1}{4} + \cdots + \frac{1}{n-2} + \frac{1}{n-1}\right)$$
$$- \left(\frac{1}{3} + \frac{1}{4} + \cdots + \frac{1}{n-2} + \frac{1}{n-1} + \frac{1}{n} + \frac{1}{n+1}\right).$$

Most of the terms cancel out (telescope), giving

$$\sum_{k=2}^{n}\frac{2}{k^2-1}=1+\frac{1}{2}-\frac{1}{n}-\frac{1}{n-1}$$
$$=\frac{3}{2}-\frac{1}{n}-\frac{1}{n+1}.$$

Since the terms $\frac{1}{n}$ and $\frac{1}{n+1}$ go to zero as $n$ goes to infinity, the series has a limiting sum of $\frac{3}{2}$.

### *1.3.2 The Harmonic Series*

The harmonic series is.

$$1+\frac{1}{2}+\frac{1}{3}+\frac{1}{4}+\frac{1}{5}+\ldots$$

We can consider the sentences of the series as.

$$1+\frac{1}{2}+\left(\frac{1}{3}+\frac{1}{4}\right)+\left(\frac{1}{5}+\frac{1}{6}+\frac{1}{7}+\frac{1}{8}\right)+\left(\frac{1}{9}+\frac{1}{10}+\cdots+\frac{1}{16}\right)+\ldots$$

Every expression is greater than or equal to the last expression in its bracket, in this case we have

$$1+\frac{1}{2}+\frac{1}{3}+\frac{1}{4}+\frac{1}{5}+\ldots$$
$$\geq 1+\frac{1}{2}+\left(\frac{1}{4}+\frac{1}{4}\right)+\left(\frac{1}{8}+\frac{1}{8}+\frac{1}{8}+\frac{1}{8}\right)+\left(\frac{1}{16}+\frac{1}{16}+\cdots+\frac{1}{16}\right)+\cdots$$
$$=1+\frac{1}{2}+\frac{1}{2}+\frac{1}{2}+\frac{1}{2}+\ldots,$$

which grows without restriction. This means that the concept of limiting sum does not apply to the harmonic series.

On the other hand, if we consider each expression as a square sentence

$$1+\frac{1}{2^2}+\frac{1}{3^2}+\frac{1}{4^2}+\frac{1}{5^2}+\ldots,$$

then this series has a finite value of $\frac{\pi^2}{6}$. This problem is known as the Basel problem states that the sum of the reciprocals of the squares of the natural numbers, $1+\frac{1}{2^2}+\frac{1}{3^2}+\frac{1}{4^2}+\frac{1}{5^2}+\ldots$, is equal to $\frac{\pi^2}{6}$. This was first proven by Leonhard Euler in 1734. The proof involves using the Taylor series expansion of $\sin(x)$ and noting that it has roots at multiples of $\pi$. By manipulating the series, Euler was able to show that the sum of the reciprocals of the squares is equal to $\frac{\pi^2}{6}$. This result has since been verified using various mathematical methods and is considered to be one of Euler's most famous contributions to mathematics.

### *1.3.3 Connection with Integration*

Integration is used to find the area under a curve. To do this, we can use the sum of the area of several rectangles to calculate the area under the curve. In fact, this method leads to a finite series. In addition, to achieve better accuracy, we use the sum of the series by taking more rectangles that have a smaller width and calculate their sum. In this case we can define the area under the curve as the limit of the sequence of sums. Note that these limits must be exist to calculate the area under the curve.

For instance, consider the function $f(x) = x^2$ on the [0, 1]. We calculate the area under $x^2$, such that the x-axis is bounded between 0 and 1.

Divide the [0, 1] into $n$ sub-intervals equal to the width $\frac{1}{n}$. On the subinterval we make two rectangles as $U_i$ and $L_i$, for $i = 1, 2, ..., n$, such that

- the height of $U_i$ is the maximum value of $f(x)$, as $\left(\frac{i}{n}\right)^2$
- the height of $L_i$ is the minimum value of $f(x)$, as $\left(\frac{i-1}{n}\right)^2$.

We can find an upper bound for the area of R by finding the area $A_n$ of $n$ rectangles $U_1, U_2, ..., U_n$ :

$$\begin{aligned} A_n &= \frac{1}{n}\left[\left(\frac{1}{n}\right)^2 + \left(\frac{2}{n}\right)^2 + \cdots + \left(\frac{n}{n}\right)^2\right] \\ &= \frac{1}{n^3}(1^2 + 2^2 + \cdots + n^2) \\ &= \frac{1}{n^3} \times \frac{1}{6}n(n+1)(2n+1) = \frac{1}{3} + \frac{1}{2n} + \frac{1}{6n^2}. \end{aligned}$$

Approximation the area under $f(x) = x^2$ on the interval is graphed in Fig. 1.5. This is an upper Riemann sum for $f$ on [0, 1].

Similarly, we can derive a lower bound for the area of $R$ by obtaining the total area $B_n$ of the $n$ rectangles as $L_1, L_2, ..., L_n$ :

$$\begin{aligned} B_n &= \frac{1}{n}\left[\left(\frac{0}{n}\right)^2 + \left(\frac{1}{n}\right)^2 + \cdots + \left(\frac{n-1}{n}\right)^2\right] \\ &= \frac{1}{n^3}(0^2 + 1^2 + \cdots + (n-1)^2) \end{aligned}$$

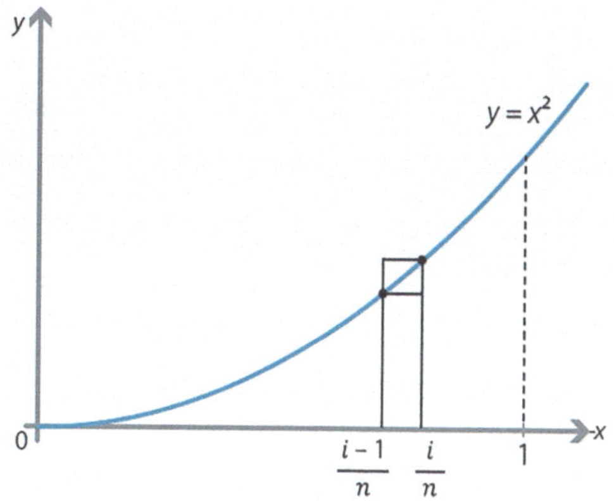


**Fig. 1.5** Approximation the area under $f(x) = x^2$

$$= \frac{1}{n^3} \times \frac{1}{6}(n-1)n(2n-1)$$
$$= \frac{1}{3} - \frac{1}{2n} + \frac{1}{6n^2}.$$

This is an lower Riemann sum for $f$ on [0, 1].
Since $B_n \leq Area R \leq A_n$, we obtain

$$\frac{1}{3} - \frac{1}{2n} + \frac{1}{6n^2} \leq Area\, R \leq \frac{1}{3} + \frac{1}{2n} + \frac{1}{6n^2}.$$

Taking the limit as $n \to \infty$, we find that the area of $R$ equals $\frac{1}{3}$. We can consider this approach as the definition of the area of $R$.

The concepts of arithmetic and geometric mean can be extended to more than two numbers.

For instance, if $a, b, c$ are positive real numbers, their arithmetic and geometric means are expressed as follows:

$$\frac{a+b+c}{3} \text{ and } \sqrt[3]{abc}$$

respectively. Generally, if $a_1, a_2, \ldots, a_n$ are $n$ positive real numbers, their arithmetic and geometric means defined as

$$\frac{a+b+c}{3} \text{ and } \sqrt[n]{a_1 a_2 \ldots a_n}$$

respectively.

The harmonic mean of two positive real numbers as $a$ and $b$ is given by

$$\frac{1}{H} = \frac{1}{2}\left(\frac{1}{a} + \frac{1}{b}\right).$$

We can rewrite this equation as follows:

$$H = \frac{2ab}{a+b}$$

The generalization of the harmonic mean for positive real numbers $a_1, a_2, ..., a_n$ can be expressed as

$$\frac{1}{H} = \frac{1}{n}\left(\frac{1}{a_1} + \frac{1}{a_2} + \cdots + \frac{1}{a_n}\right).$$

### 1.3.4 *The AM–GM Inequality*

Exercise 11 gave a geometric proof that the arithmetic mean of two positive numbers a and b is greater than or equal to their geometric mean. We can also prove this algebraically, as follows.

According to the real number $a$ and $b$ are positive, we define $x = \sqrt{a}$ and $y = \sqrt{b}$. In this case

$$(x-y)^2 \geq 0 \Rightarrow x^2 + y^2 - 2xy \geq 0$$
$$\Rightarrow \frac{x^2 + y^2}{2} \geq xy$$

and so

$$\frac{a+b}{2} \geq \sqrt{ab}$$

The above statement is called AM-GM inequality. We note the AM-GM inequality becomes equality if and only if $a = b$.

***Example*** Derive the range of the function $f(x) = x^2 + \frac{1}{x^2}$, for $x \neq 0$.

*Solution* Applying the AM–GM inequality,

$$f(x) = x^2 + \frac{1}{x^2} \geq 2\sqrt{x^2 \times \frac{1}{x^2}} = 2.$$

Therefore, the range of $f$ exists in the $[2, \infty)$. On the other hand, $f(1) = 2$ and $f(x) \to \infty$ as $x \to \infty$. Since $f$ is a continuous function, for each $y \geq 2$, there is $x \geq 1$ with $f(x) = y$. Thus, $[2, \infty)$. is the range of $f$.

The AM–GM inequality can be generalized as follows. If $a_1, a_2, ...a_n$ are $n$ positive real numbers, then

$$\frac{a_1 + a_2 + \cdots + a_n}{n} \geq \sqrt[n]{a_1 a_2 ... a_n}.$$

Exercise 11 proves this result.

## 1.3.5 History and Applications

### 1.3.5.1 An Application to Film and Video

In editing a film, one has the ability to change the width and height of the images being shot. The ratio of the width and height of the image is called the ratio of the dimensions of the film. This scale is usually expressed in $x : y$ format.

The aspect ratio as $2.35 : 1$ is very common in movie theatres. While the aspect ratio traditionally used for TV and movies is $4 : 3 \approx 1.33 : 1$. It was found that the geometric mean of the numbers 2.35 and 1.33 creates a good fit between two proportions of different dimensions and distorts or cuts them both in the same way. The geometric mean of 2.35 and 1.33 is approximately 1.77 and the $1.77 : 1$ ratio is approximately $16 : 9$. This is a dimensional ratio approved by the Association of Film and Television Engineers and used for high quality digital TVs.

### 1.3.5.2 Fibonacci Numbers

Consider the Fibonacci sequence as follows [1]:

$$1, 1, 2, 3, 5, 8, 13, 21, 34, 55, 89, 144, ...$$

In the Fibonacci sequence, each expression is obtained from the sum of the previous two expressions. Thus if $a_n$ is $n$th sentence, we have.

$a_1 = a_2 = 1$ and $a_n = a_{n-1} + a_{n-2}$, for $n \geq 3$.

This problem represents a second-order linear recurrence relation model.

Consider a first-order linear recurrence model as $u_n = ku_{n-1}$, where $k$ is a constant. In this case, we have

$$a_n = a_{n-1} + a_{n-2}$$

and

$$\begin{aligned} a_n - ka_{n-1} &= a_{n-1} + a_{n-2} - ka_n \\ &= (1-k)a_{n-1} + a_{n-2} \\ &= (1-k)\left(a_n + \frac{1}{1-k}a_{n-2}\right) \end{aligned}$$

$$\Rightarrow \frac{1}{1-k} = -k.$$

Thus,

$$k^2 - k - 1 = 0,$$

and $k = \frac{1}{2}\left(1 \pm \sqrt{5}\right)$. Therefore, we obtain
$a_n = A_1 \times \left(\frac{1+\sqrt{5}}{2}\right)^n$ and $a_n = A_2 \times \left(\frac{1-\sqrt{5}}{2}\right)^n$.

According to the recurrence relations theory the general solution of this problem is provided by summing these two solutions as follows:

$$a_n = A_1 \times \left(\frac{1+\sqrt{5}}{2}\right)^n + A_2 \times \left(\frac{1-\sqrt{5}}{2}\right)^n.$$

Applying the initial condition $a_1 = a_2 = 1$ gives $A_1 = \frac{1}{\sqrt{5}}$ and $A_2 = \frac{1}{\sqrt{5}}$. Finally, the $n$th term of the Fibonacci sequence is:

$$a_n = \frac{1}{\sqrt{5}}\left(\left(\frac{1+\sqrt{5}}{2}\right)^n - \left(\frac{1-\sqrt{5}}{2}\right)^n\right).$$

Notice that $\frac{1}{2}\left(1 + \sqrt{5}\right)$ which appears in this problem is the **golden ratio**. One of the interesting facts about the Fibonacci sequence is that 1 and 144 are the only perfect squares.

### 1.3.6 *The Greeks*

Ancient Greek mathematicians had a special interest in ratios. In fact, many of their calculations are done geometrically using length and ratio. They found the connection between music and ratio, and this is the main reason for naming the harmonic mean.

The Greeks defined several different means. Now we will deal with the Heronian mean, which is defined for two positive real numbers a and b, and is denoted by $N(a, b)$. This mean is the average of the three numbers a, b and their geometric mean. In other words,

$$N(a, b) = \frac{a + b + \sqrt{ab}}{3}.$$

Notice that, for $a, b > 0$, we have

$$\frac{a+b+\sqrt{ab}}{3} \geq \frac{2\sqrt{ab}+\sqrt{ab}}{3} = \sqrt{ab}$$

and

$$\frac{a+b+\sqrt{ab}}{3} \leq \frac{2\sqrt{ab}+\frac{a+b}{2}}{3} = \frac{a+b}{2},$$

giving

$$\sqrt{ab} \leq N(a,b) \leq \frac{a+b}{2}.$$

The Heronian mean is between the arithmetic and geometric means.

Generalizing the Heronian mean for $n$ positive real numbers $a_1, a_2, ..., a_n$, we obtain

$$N = \frac{a_1 + a_2 + \cdots + a_n + \sqrt[n]{a_1 a_2 ... a_n}}{n+1}.$$

The Greeks were concerned in different types of series and sequences. They also knew the square numbers sequence as

$$1, 4, 9, 16, ...$$

They also introduced the triangular numbers sequence

$$1,\ 3,\ 6,\ 10,\ ...$$

given by $T = 1$ and $T_n = T_{n-1} + n$, for $n \geq 2$. The general term of this sequence is

$$T_n = 1 + 2 + \cdots + n = \frac{1}{2}n(n+1).$$

The graphic representation of these numbers can be shown as follows, which represent the first four triangular numbers: (Fig. 1.6).

***Example*** Show that

**Fig. 1.6** First four triangular numbers

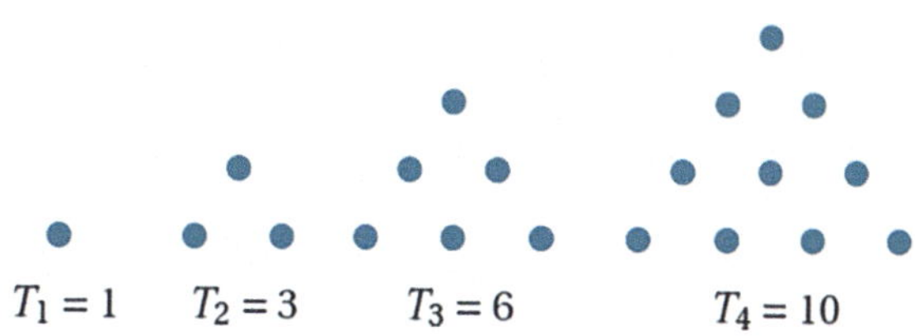

I. $1^2 + 2^2 + 3^2 + \cdots + n^2 = \frac{1}{6}n(n+1)(2n+1)$
II. $1^3 + 2^3 + 3^3 + \cdots + n^3 = \frac{1}{4}n^2(n+1)^2$.

*Solution*

I. For the general case, $n$ terms are to be summed

$$S = a_0 + a_1 + a_2 + \cdots + a_{n-1}.$$

The formula for the $n$th term is

$$a_n = a_0 + (a_1 - a_0)n + (a_2 - 2a_1 + a_0)\frac{n(n-1)}{2}.$$

Using the results

$$\sum_{i=0}^{n-1} i = \frac{n(n-1)}{2}$$

And

$$\sum_{i=0}^{n-1} i(i-1) = \frac{n(n-1)(n-2)}{3}$$

leads to

$$S = \sum_{i=0}^{n-1} a_i = a_0 n + (a_1 - a_0)\frac{n(n-1)}{2} + (a_2 - 2a_1 + a_0)\frac{n(n-1)(n-2)}{6}.$$

II. Define

$$A(n) = \sum_{i=1}^{n} i \quad \text{and} \quad B(n) = \sum_{i=1}^{n} i^2 .$$

Then

$$A(n) = \frac{n(n+1)}{2} \text{ and } B(n) = \frac{n(n+1)(2n+1)}{6}.$$

Let $S(n) = \sum_{i=1}^{n} i^3$. Then

$$\begin{aligned} S(n) &= n^3 + (n-1)^3 + (n-2)^3 + \cdots + 2^3 + 1^3 \\ &= \sum_{k=0}^{n-1} (n-k)^3 \end{aligned}$$

$$= \sum_{k=0}^{n-1} \left(n^3 - 3n^2k + 3nk^2 - k^3\right)$$
$$= n^4 - 3n^2.A(n-1) + 3.B(n-1) - S(n-1).$$

In this case,

$$S(n) = n^4 - 3n^2.\frac{n(n-1)}{2} + 3n.\frac{n(n-1)(2n-1)}{6} - S(n-1)$$

And

$$S(n) = n^3 + S(n-1).$$

Thus,

$$2S(n) = n^4 + n^3 - 3n^2.\frac{n(n-1)}{2} + 3n.\frac{n(n-1)(2n-1)}{6}$$
$$= \frac{n^2(n+1)^2}{2}.$$

Therefore,

$$S(n) = \frac{n^2(n+1)^2}{4}.$$

## 1.4 Integer Divisibility

In this section, the divisibility of integers $\mathbb{Z}$ is discussed. We provide detailed evidence on some of the basic properties about divisibility. In fact, these properties are quite evident, but proving them can give us good ideas for solving problems.

**Definition** An integer $b \neq 0$ divides another integer $a$ if and only if $\exists k \in \mathbb{Z}$ that $a = kb$.

We also say that $b$ is a factor (or divisor) of $a$. One frequently writes $b|a$ to indicate that $b$ divides $a$.

***Example*** $3|12$: 3 divides 12 but $5 \not| 12$: 5 does not divide 12.

***Example*** Let a and b be two positive integers in $\mathbb{Z}^+$ such that $a > b$. How many positive integers not exceeding $a$ are divisible by $b$? In other words, find such $c \in \mathbb{Z}^+$ that $b < c < a$ and $b|c$.

*Solution* All numbers divisible by $b$ are in the form $b \in k$ where be $k \in \mathbb{N}^+$. They are positive and do not exceed $a$,

$$a < b \times k \leq a or 0 < k \leq \frac{a}{b}$$

Therefore, there are $floor(\frac{a}{b})$ or $\lfloor \frac{a}{b} \rfloor$ such integers.

**Theorem** *For all integers $a, b, c \in \mathbb{Z}$.*

1. $1|a$, $-1|a$ *and* $a|0$.
2. *Reflexivity:* $a|a$.
3. *Transitivity:* $a|b \wedge b|c \Rightarrow a|c$.
4. *Not-quite antisymmetry:* $a|b \wedge b|a \Rightarrow a = b \vee c = -b$.
5. *If* $a|b \wedge a|c \Rightarrow a|(n \times b + m \times c)$, *for any integers n and m.*

***Proof***

(1) and (2) follow immediately

$a \in \mathbb{Z} \Rightarrow a = a \times 1 \Rightarrow a|a$.

(3) Given $b = x \times a$ and $c = y \times b$. Then $c = y \times b = y \times x = (x \times y) \times a$, so $a|c$.
(4) Given $b = x \times a$ and $a = y \times b$. Then $a = xya \Rightarrow a(1 - xy) = 0$, and therefore $xy = 1$ (there are no zero-divisors in the integers). It follows that either $x = y = 1$ or $y = x = -1$. But $x = -1$ implies $a = b$, and $x = -1$ implies $b = -a$.
(5) Given $b = x \times a$ and $c = y \times a$. Consider $n \times b + m \times c$

$nb + mc = xan + yam = a(xn + ym) \Rightarrow a|(nb + mc)$.
It follows $a|(nb + mc)$.

Do there exist integers $x$, $y$, and $z$ such that $6x + 9y + 15z$? No, they don't. Here is the proof by contradiction. Since 3, 6, and 9 have a common divisor 3, then 3 must divide its linear combination $3|(6x + 9y + 15z) = 3|107$, which is wrong.

### *1.4.1 Primes*

Observation. Every positive integer has at least two divisors: 1 and itself.

**Definition** An integer $p_1$ is called a prime if its only positive divisors are 1 and p. Otherwise it is called a composite.

The number 1 is a special case which is considered neither prime nor composite. The number 2 is also special; it is the only even prime.

**Theorem** *There are an infinite number of primes.*

***Proof*** (by contradiction)

Assume otherwise, saying $p_1, \ldots, p_n$ is a complete list of all primes.

Define

$$p = (p_1 \times p_2 \times \ldots \times p_n) + 1.$$

Since this number $p$ is bigger than all the $p_i$, it is not a prime according to the assumption.

But then, there is some prime that divides $p$. Since our list is supposedly complete that prime must be, say, $p_r$.

We have that $p_r|p$.

$$\begin{cases} p_r|p \\ p_r|(p_1.p_2 \ldots p_n) \end{cases} \Rightarrow p_r|(p - p_1p_2 \ldots p_n) \Rightarrow p_r|1.$$

But then $p_r = 1$.

A contradiction.

QED. End of proof ("quod erat demonstrandum").

#### 1.4.1.1 Goldbach Conjecture (1742)

Prime numbers satisfy many strange and wonderful properties.

Observation:

$$\begin{aligned} 6 &= 3 + 3 \\ 7 &= 5 + 2 \\ 8 &= 5 + 3 \\ 9 &= 7 + 2 \\ 10 &= 7 + 3 \\ 18 &= 11 + 7 \end{aligned}$$

What about the number 117? Can you show it as sum of two primes?

Goldbach speculated that any number greater than 5 could be written as the sum of the three prime numbers.

Euler answered Goldbach's theory by stating that any even number is greater than twice the sum of the first two numbers.

$$p_1 + p_2 = 2n.$$

This hypothesis was confirmed numerically in 2003 for numbers up to $6 \times 10^{16}$

#### 1.4.1.2 Mersenne Numbers

For some years, people believed that if $p$ is prime, then so is $2^p - 1$ :

$$2^2 - 1, 2^3 - 1, 2^5 - 1, \cdots .$$

This is not true for all primes; for example

$$2^{11} - 1 = 2047 = 23 \times 89.$$

**Mersenne Conjecture** and composite for all other p = 2,3,5,7,13,17,19,31,127 prime number for $2^p - 1$ positive intergers.

**Theorem** *If $2^p - 1$ is prime, then p is prime.*

***Proof*** (by contradiction) We assume that $2^p - 1$ is prime, but p is not prime. Let $p$ be a composite number, $p = r \times s$. Consider the following polynomial

$$x^{r \times s} - 1$$

It can be written as

$$x^{r \times s} - 1 = \left(x^s - 1\right)\left(x^{s(r-1)} + x^{s(r-2)} + ...x^s + 1\right),$$

which is easily proved by expanding the right-hand side. Therefore, if $p$ is composite, then $x^p - 1$ is composite, so is $2^p - 1$, since it's divisible by $2^s - 1$. Contradiction to our assumption that $2^p - 1$ is prime. QED.

### *1.4.2 Perfect Numbers*

**Definition** A positive integer is a perfect number if it equals the sum of its proper divisors (not including itself).

The first few perfect numbers are

$$\begin{aligned}
&6 = 1 + 2 + 3 \\
&28 = 1 + 2 + 4 + 7 + 14 \\
&496 = 1 + 2 + 4 + 8 + 26 + 32 + 62 + 24 + 248 \\
&8128 = ...
\end{aligned}$$

***Question*** Are they all even?

This question is much, much harder. It is not known if any odd perfect numbers exist.

All even perfect numbers are in the form

$$\begin{aligned}
6 &= 1 + 2 + 3 = 2 \times 3 \\
28 &= 1 + 2 + 3 + 4 + 5 + 6 + 7 = 4 \times 7 \\
496 &= 1 + 2 + 3 + \cdots + 31 = 16 \times 31 \\
8128 &= 1 + 2 + 3 + \cdots + 127 = 64 \times 127.
\end{aligned}$$

Generally, $2^{n-1}(2^n - 1)$, when n is prime.

This is a relation between the perfect and the Mersenne primes. So the search for Mersenne primes is also the search for even perfect numbers!

### 1.4.3 The Division Algorithm

The division algorithm is not really an algorithm in the sense of the word computer science, but is a claim that can be made from the integer division.

**Theorem** (The Division Algorithm). *Let b > 0, and a arbitrary integers. Then, there exist and unique integers q (the quotient) and r (the remainder) such that*

$a = q \times b + r$, *where* $0 \leq r < b$.

*Dividend* = *Quotient* × *Divisor* + *Remainder.*

***Proof*** First we prove **existence**. Two cases:

(a) b does divide a.
(b) b does not divide a.

*Case a)* is trivial.

*Case b)* Given that b does not divide a. For example, if $b = 3$ and $a = 10$, consider the following set

$$\begin{aligned}
\{10 - 3q\} &= \{1, 4, 7, 10, 13, \ldots\} \\
q &= 3, 2, 1, 0, -1, \ldots
\end{aligned}$$

$S$ is not empty. We must prove this statement. Again, two cases to consider.

If a > 0, choose $q = 0$, so $a \in S$.

If $a < 0$, choose $q =$ a, then $a - ab = -a(b - 1) > 0$ is in the set. Observe that $a(b - 1) \neq 0$ since b cannot be equal 1. Why? Because of our assumption "$b$ does not divide $a$."

**Axiom** (Well-Ordering Principle)

Each set of natural numbers contains the smallest element.

Since $S$ is a set of non-negative integers, it must have a minimum term such as $r_0$.

$$r_0 = \min(S)$$
$$0 < r_0 = a - q_0 \times b.$$

We prove that $r_0 < b$ (by contradiction).
**Case 1.** Let $r_0 = b$
Substitute this into the above formula for $r_0$

$$0 < b = a - q_0 b.$$

Collect terms in b

$$a = bq_0 b = (q_0 + 1)b.$$

This implies that $b|a$, which contradicts to Case b) assumption that $b \not| a.b \not| a.$
**Case 2.** Let $r_0 > b > 0$. Consider $r_0 - b$. It is positive $r_0 - b > 0$ and also $r_0 - b < r_0$. Next we show that $r_0 - b \in S$

$$r_0 - b = (a - bq_0) - b = a - q_0(q_0 + 1) \in S.$$

Therefore, $r_0 - b$ is in S and it is smaller than $r_0$. Contradiction to the minimality of $r_0.r_0$.
It remains to prove that q and r are **unique**.
Suppose that

$$a = bq_0 + r_0$$
$$a = bq_1 + r_1$$

where $0 \leq r_i < b$ and $r_0 \leq r_1$.
Subtract one equation from another, we obtain

$$0 = bq_0 - bq_1 + r_0 - r_1,$$
$$0 = b(q_0 - q_1) + r_0 - r_1,$$
$$b(q_0 - q_1) = r_1 - r_0.$$

According to our assumption $r_0 - b$. Therefore,

$$0 \leq b(q_0 - q_1) < b$$
$$0 \leq (q_0 - q_1) < 1$$
$$q_0 = q_1.$$

**The cancellation law**. We have used an important property of the integers: $x \times y = 0$ implies $x = 0$ or $y = 0$.
It says that there are no nonzero zero-divisors in the integers.

Here is a simple application of the Division Algorithm.

**Lemma** *Let p be prime. Then* $p|(a \times b)$ *implies that* $p|a$ *or* $p|b$.

***Proof*** From the division algorithm, we have

$$a = q_1 \times p + r_1 and b = q_2 \times p + r_2.$$

where $0 \leq r_1, r_2 < p$. Therefore,

$$a \times b = q_1 q_2 p^2 + q_1 r_2 p + q_2 r_1 p + r_1 r_2$$
$$a \times b = p(q_1 q_2 p + q_1 r_2 + q_2 r_1) + r_1 r_2$$

According to $p$ divides $a \times b = 0$, therefore, $p|(r_1 \times r_2)$. It follows then that the remainder $r_1 \times r_2$ must be 0. But then $r_1$ or $r_2$ must be 0, so that $p$ divides a or $b$. QED.

**Application**. We prove that $\sqrt{2}$ is irrational.

***Proof*** (by contradiction). Let $\sqrt{2} = \frac{p}{q}$ Then

$$\sqrt{2} \in \mathbb{Q} \Rightarrow \exists p, q \in \mathbb{Z} : \sqrt{2} = \frac{p}{q}, \gcd(p, q) = 1$$

$$\Rightarrow 2 = \frac{p^2}{q^2} \Rightarrow 2q^2 = p^2 \Rightarrow 2|p^2 \Rightarrow 2|(p \times q) \Rightarrow 2|p.$$

Assume $p = 2c,\ c \in \mathbb{Z}^+$. Then

$$2q^2 = p^2 \Rightarrow 2q^2 = 4c^2 \Rightarrow q^2 = 2c^2 \Rightarrow 2|q^2 \Rightarrow 2|q \Rightarrow \gcd(p, q) \geq 2.$$

Contradiction, p and q have a common divisor 2.

### *1.4.4 The Fundamental Theorem of Arithmetic*

An important feature of prime numbers is that any positive integer can be expressed as the product of prime numbers.

**Theorem** *(The Fundamental Theorem of Arithmetic). Let* $n \geq 2$ *be an integer. Then there exist primes* $p_1, p_2, ..., p_k$ *such that*

$$n = p_1 \times p_2 \times \cdots \times p_k.$$

*If we require in addition that the sequence $p_1, p_2, ..., p_k$ is ordered, then it is uniquely determined by n.*

*Example of prime factorization*

$$300 = 2 \times 2 \times 3 \times 5 \times 5.$$

***Proof*** First existence. We use in induction on $n$.

The base case $n = p_1 \times p_2 \times \cdots \times p_k$. is obvious.

Inductive Hypothesis: Numbers up to $n - 1$ can be written as a product of primes.

Inductive step: If $n$ is prime, there is nothing to show.

Otherwise, $n = a \times b$ where $1 < a, b < n$.

By IH (inductive hypothesis), both $a$ and $b$ can be written as products of primes, and our claim follows.

Uniqueness of the factorization. Suppose

$$n = p_1 \times p_2 \times \cdots \times p_k = q_1 \times q_2 \times \cdots \times q_s.$$

That is,

$$n = \prod_{i=1}^{k} p_i = \prod_{i=1}^{6} p_i,$$

Due to the fact that $p_1$ divides the second product, we have $p_1 | q_i$ for some $i$. Thus, $p_1 = q_i$. Via a similar argument, we have $q_1 = p_j$ for some j.

It follows from the order assumption that $q_1 = p_j$ so that

$$p_2 \times \cdots \times p_k = q_2 \times \cdots \times q_s.$$

By continuing in this way, we see that each $p_k$ must be paired with $q_j$, apart from the order of the factors. QED.

**Corollary** *Every positive integer > 1 can be written uniquely (except for order) in the form*

$$x = p_1^{e_1} \times p_2^{e_2} \times \cdots \times p_n^{e_m}, \ p_1 \neq p_1 fori \neq j.$$

### 1.4.5 Counting Divisors

Observation. Count the number of positive divisors

- 2 has two divisors 1 and 2,

$$2^2 = 2^1$$

- 4 has three divisors 1, 2 and 4,

$$4 = 2^2$$

- 12 has six divisors 1, 2, 3, 4, 6 and 12,

$$12 = 2^2 \times 3$$

- 300 has 18 divisors,

$$300 = 2^2 \times 3 \times 5^2.$$

Each divisor of $300 = 2^2 \times 3 \times 5^2$ must be in the form

$$2^i \times 3^j \times 5^k$$

where $0 \le i \le 2, 0 \le j \le 1, 0 \le k \le 2$, otherwise it won't divide 300. These will give us $3 \times 2 \times 3 = 18$ choices. We use here the rule of product.

#### 1.4.5.1 The Rule of Product

Suppose in one day you have 5 choices about going to the cinema and 3 choices about going to a party. In this case, how many ways can you spend that day?

**Theorem** *The following integer*

$$x = p_1^{e_1} \times p_2^{e_2} \times \cdots \times p_n^{e_m}$$

*has* $(e_1 + 1) \times (e_2 + 1) \times \cdots \times (e_m + 1)$ *divisors.*

***Example*** 72 has 12 divisors:

$$72 = 2^3 \times 3^2.$$

The positive divisors of an integer are pictured in a Hasse diagram as follows: two divisors $a$ and $b$ are connected by a path going up if $a$ is a divisor of $b$ (Fig. 1.7).

How does the diagram help us to do algebra? It can be used to compute GCD and LCM. Observe that each element in the diagram generates a downward cone of divisors and upward code of multiples. Therefore, the intersection of downward cones gives the GCD of two numbers, and the intersection of two upward cones gives the LCM.

**Fig. 1.7** The positive divisors of an integer in a Hasse diagram

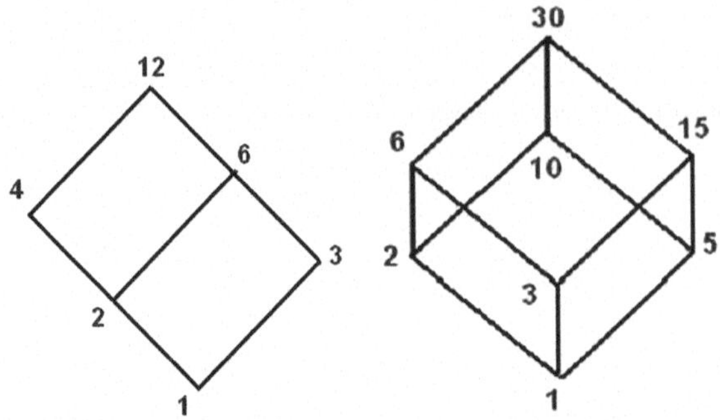


**Problem** A football team has 100 lockers that are initially closed. Student #1 opens all lockers. Student #2 closes all even lockers. Student #3 changes the status of each locker that is a multiple of 3, and so on..., student #$k$ changes the status of each locker that is a multiple of $k$. Which lockers are open after all 100 students have done their job?

*Solution* Think about what it takes to make a locker to end up open. It takes an odd number of changes. Which means an odd number of divisors.

Integer $x = p_1^{e_1} \times p_2^{e_2} \times \cdots \times p_n^{e_m}$ divisors. Thus $(e_1+1) \times (e_2+1) \times \cdots \times (e_m+1)$ is odd or each factor is odd. Therefore, each $(e_1 + 1) \times (e_2 + 1) \times \cdots \times (e_m + 1)$ is even. So x must be a perfect square. Open lockers are those which are $1, 4, 9, 16, 25, \ldots$

## 1.4.6 Greatest Common Divisor

**Definition** Suppose $a$ and $b$ are integers. A greatest common divisor (GCD) is defined as

$$\gcd(a, b) = \max(d \in P|d|a \wedge d|b).$$

**Definition** Integers a and b are relatively prime (or coprime) if and only if $\gcd(a, b) = 1$.

It is clear from the definition that

$$\gcd(-a, -b) = \gcd(a, b)$$
$$1 \leq \gcd(a, b) \leq \min(a, b).$$

Given a pair of numbers. Does a GCD always exist? Is it unique?

Let us prove it for positive integers!

**heorem** $\forall a, b \in \mathbb{N}^+, \exists! d \in \mathbb{Z}$, that is the GCD of a and b.

***Proof*** Let

$$S = \{n \times a + m \times bn, m \in \mathbb{Z}, n \times a + m \times b > 0\}.$$

Set $S$ has a least element. Let us call it $d$. We need to prove that $d$ is a GCD.
First we prove that d exists; in other words, that $d|a$ and $d|b$.
By contradiction (on minimality of $d$).
Assume that $d \not| a$. Using the division algorithm, $a = qd + r$ where $0 < r < d$.

$$r = a - qd = a - q(na + mb) = (1 - qn)a + (-qm)b.$$

We see that $r \in S$ and $r < d$. Contradiction, on minimality of d.

Next we prove that $d$ is a greatest. Assume that there is c such that $c|a$ and $c|b$. Therefore, $c$ divides their linear combination $c|(n \times a + m \times b) \Rightarrow c|d$.

Finally we prove that $d$ is unique. Let $d_1$ and $d_2$ be GCDs of $a$ and $b$. Suppose $d_1$ is a greatest, and $d_2$ is a common divisor. Then from the previous paragraph, we see that $d_2|d_1$. Reversing rules, then $d_2|d_1$. Therefore, $d_1|d_2$. QED.

**Corollary**

$$\gcd(a, b) = n \times a + m \times b, wheren, m \in \mathbb{Z}.$$

*For example,*

$$\gcd(56, 24) = 8 = 56 \times 1 + 24 \times (-2).$$

### *1.4.7 Euclidean Algorithm*

Given a pair of numbers $a$ and $b$,
find $\gcd(a, b)$.
Find such n, m that $\gcd(a, b) = n \times a + m \times b$
Here is a useful lemma that is used in an algorithm for finding GCD:

**Lemma** *Let* $a = bq + r$, *then* $\gcd(a, b) = \gcd(b, r)$.

***Proof*** Denote

$$d_1 = \gcd(a, b)$$
$$d_2 = \gcd(b, r).$$

Since $d_1 = \gcd(a, b) \Rightarrow d_1|r \Rightarrow d_1$ divides both $r$ and $b$, therefore $d_1|d_2$.

Conversely, since $d_2 = \gcd(b, r) \Rightarrow d_2|a \Rightarrow d_2$ divides both $a$ and $b$, therefore $d_2|d_1$. Hence, $d_1 = d_2$. QED.

**heorem** (Euclidean algorithm). *The algorithm computes a GCD of two positive integers a, b. We apply the division algorithm*

$$\begin{aligned}
a &= b \times q_1 + r_1, & 0 \le r_1 < b \\
b &= b \times q_1 + r_1, & 0 \le r_2 < r_1 \\
r_1 &= r_2 \times q_3 + r_3, & 0 \le r_3 < r_2 \\
& & \vdots \\
r_{k-2} &= r_{k-1} \times q_k + r_k, & 0 \le r_k < r_{k-1} \\
r_{k-1} &= r_k \times q_{k+1} + 0.
\end{aligned}$$

*The last non-zero remainder is GCD:* $\gcd(a, b) = r_k$.

***Proof*** Let

$$S = \{r_1, r_2, ..., r_k\}.$$

Set $S$ contains a least element. This means that the above division will definitely terminate. Next

$$\gcd(a, b) = \gcd(b, r_1) = \gcd(r_1, r_2) = \cdots = \gcd(r_{k-1}, r_k) = \gcd(r_k, 0) = r_k.$$

QED.

***Question*** When does the worst performance of the Euclidean algorithm occur?
Let us run the algorithm for gcd (21, 13)

$$\begin{aligned}
21 &= 1 \times 13 + 8 \\
13 &= 1 \times 8 + 5 \\
8 &= 1 \times 5 + 3 \\
5 &= 1 \times 3 + 2 \\
3 &= 1 \times 2 + 1 \\
2 &= 2 \times 1 + 0.
\end{aligned}$$

These are Fibonacci numbers (they usually denote as $F_k$)

$$1, 1, 2, 3, 5, 8, 13, 21, 34, ...$$

Therefore, if we set all quotients $q_i$ to 1 in the Euclidean algorithm, we immediately obtain the following equation for the remainders

$$r_{k-2} = r_{k-1} + r_k$$

which defines Fibonacci numbers backwards. We will talk about these numbers in more details later in the course.

**heorem** *The Euclidean algorithm computes* $\gcd(F_{k-1}, F_k)$ *in* $k - 1$ *steps.*

**heorem** *For two positive m and $n \leq m$ the Euclidean algorithm computes* $\gcd(m, n)$ *in no more that* $\log \phi + 1$ *steps.*

Here the value $\phi$, phi, is the golden ratio defined by $\phi = \frac{\sqrt{5}+1}{2}$ as a solution of $x^2 - x - 1 = 0$.

***Example*** Let run the Euclidean algorithm to find gcd(203, 91).

$$203 = 2 \times 91 + 21$$
$$91 = 4 \times 21 + 7$$
$$21 = 3 \times 7 + 0.$$

In the following, we reverse these equations. Based on the first equation we have

$$21 = 203 - 2 \times 91.$$

The second equation gives

$$7 = 91 - 4 \times 21.$$

We substitute the first equation into the second.

The method we followed above is a bit confusing because of all the substitutions we have to make. By performing some auxiliary calculations in the Euclidean algorithm, the amount of computation involved in finding $n$ and $m$ can be reduced (and no replacement will be required). This method is known as the extended Euclidean algorithm.

The extended Euclidean algorithm proceeds as follows. In accordance with the Euclidean algorithm

$$\begin{aligned}
a &= b \times q_1 + r_1, & r_1 &= a + b(-q_1)\\
b &= b \times q_1 + r_1, & r_2 &= b + r_1(-q_2)\\
r_1 &= r_2 \times q_3 + r_3, & r_3 &= r_1 + r_2(-q_3)\\
& & &\vdots\\
r_{k-2} &= r_{k-1} \times q_k + r_k, & r_k &= r_{k-2} + r_{k-1}(-q_k)\\
r_{k-1} &= r_k \times q_{k+1} + 0.
\end{aligned}$$

we build the matrix

$$\begin{pmatrix} a & 1 & 0 \\ b & 0 & 1 \\ r_1 & 0 & 1 \\ r_2 & -q_2 & 1+q_1q_2 \\ & \vdots & \\ r_k & n & m \end{pmatrix} \begin{matrix} \times(-q_1) \\ \times(-q_2) \\ \times(-q_3) \\ \vdots \end{matrix}$$

where the third row is the first row minus the second row multiplied by $-q_1$ and so on.

***Example*** Use the algorithm to find gcd(123, 45).

$$\begin{aligned} 123 &= 2 \times 45 + 33 \\ 45 &= 1 \times 33 + 12 \\ 33 &= 2 \times 12 + 9 \\ 12 &= 1 \times 9 + 3 \\ 9 &= 3 \times 3 + 0. \end{aligned}$$

On the other hand, the matrix for n and m:

$$\begin{pmatrix} 123 & 1 & 0 \\ 45 & 0 & 1 \\ 33 & 1 & -2 \\ 12 & -1 & 3 \\ 9 & 3 & -8 \\ 3 & -4 & 12 \end{pmatrix} \begin{matrix} \\ \times(-2) \\ \times(-1) \\ \times(-2) \\ \times(-1) \\ \\ \end{matrix}$$

Therefore,

$$\gcd(123, 45) = (-4) \times 123 + 11 \times 45.$$

#### 1.4.7.1 Applications

Remember the movie Die Hard with a Vengeance? Bruce Willis and Samuel L. Jackson are set to disarm a bomb by accurately measuring 4 gallons of water using 3- and 5-gallon containers. In this section we describe the mathematics used to solve these types of problems.

*Problem* You have two containers that hold 17 and 55 oz (oz). How would you measure 1 oz?

*Solution* Apply the Euclidean algorithm, find the GCD, and then consider it as a linear combination of 17 and 55.

$$55 = 17 \times 3 + 4$$
$$17 = 4 \times 4 + 1$$

$$\gcd(55, 17) = 13 \times 17 - 4 \times 55.$$

In fact, a smaller container should be filled 12 times $12 \times 17 = 204$ and then emptied into a larger container with capacity

$$12 \times 17 = 204 = 33 \times 55 + 39.$$

At this point, a smaller empty container and 39oz in a larger container are available. In the next step (thirteenth) 16oz is filled in a larger container and 1oz is placed in a smaller container.

**Problem** Suppose someone makes every MATLAB program in 10 min and every C program in 6 min. If he (she) works for 104 min, how many programs can he fix?

*Solution* Write an equation

$$10 \times x + 6 \times y = 104 \text{ } or \text{ } 5 \times x + 3 \times y = 52.$$

Since

$$\gcd(5, 3) = 1 = 2 \times 3 - 1 \times 5.$$

Multiply it by 52

$$52 = 104 \times 3 - 52 \times 5.$$

Add and subtract $15k$, where $k \in \mathbb{Z}$ and

$$52 = (104 - 5k) \times 3 + (3k - 52) \times 5.$$

We choose $k$ in a such manner that

$$\begin{cases} 104 - 5k > 0 \\ 3k - 52 > 0 \end{cases} \Rightarrow \frac{52}{3} < k < \frac{104}{5}.$$

This gives us three possible solutions $k = 18, 19, 20$. Therefore, he (she) can fix (14 MATLAB and 2 C) or (9 MATLAB and 5 C) or (4 MATLAB and 8 C) programs.

### 1.4.8 Continued Fractions

In mathematics, a continued fraction is an expression obtained through an iterative process of representing a number as the sum of its integer part and the reciprocal of another number, then writing this other number as the sum of its integer part and another reciprocal, and so on. In a finite continued fraction (or terminated continued fraction), the iteration/recursion is terminated after finitely many steps by using an integer in lieu of another continued fraction. In contrast, an infinite continued fraction is an infinite expression. In either case, all integers in the sequence, other than the first, must be positive. Let $b > 0$, and $a$ arbitrary integers. Then, from division algorithm, we have

$$\begin{aligned}
a &= b \times q_1 + r_1, & a &\le r_1 < b\\
b &= r_1 \times q_2 + r_2, & a &\le r_2 < r_1\\
r_1 &= r_2 \times q_3 + r_3, & a &\le r_3 < r_2\\
&\vdots\\
r_{k-2} &= r_{k-1} \times q_k + r_k, & a &\le r_k < r_{k-1}\\
r_{k-1} &= r_k \times q_{k+1} + 0.
\end{aligned}$$

From the first line we express $\frac{a}{b}$

$$\frac{a}{b} = q_1 + \frac{r_1}{b} \Rightarrow \frac{a}{b} = q_1 + \frac{1}{\frac{b}{r_1}}$$

From the second line we express $\frac{b}{r_1}$

$$\frac{b}{r_1} = q_2 + \frac{r_2}{r_1} \Rightarrow \frac{b}{r_1} = q_2 + \frac{1}{\frac{r_1}{r_2}}.$$

Substitute $\frac{b}{r_1}$ into the equation for $\frac{a}{b}$

$$\frac{a}{b} = q_1 + \frac{1}{q_2 + \frac{1}{\frac{r_1}{r_2}}}.$$

One more iteration

$$\frac{a}{b} = q_1 + \frac{1}{q_2 + \frac{1}{q_3 + \frac{1}{\frac{r_2}{r_3}}}}.$$

Generally,

$$\begin{aligned}
\frac{a}{b} &= q_1 + \frac{r_1}{b} \\
\frac{b}{r_1} &= q_2 + \frac{r_2}{r_1} \\
\frac{r_1}{r_2} &= q_3 + \frac{r_3}{r_2} \\
&\vdots \\
\frac{r_{k-2}}{r_{k-1}} &= q_k + \frac{r_k}{r_{k-1}} \\
\frac{r_1}{r_1} &= q_1 + 0.
\end{aligned}$$

We will get

$$\frac{a}{b} = q_1 + \frac{1}{q_2 + \frac{1}{q_3 + \frac{1}{q_4 + \cdots + \frac{1}{q_{k+1}}}}}.$$

This expression is called a continued fraction for a rational number. Examples,

$$\frac{37}{11} = 3 + \frac{1}{2 + \frac{1}{1 + \frac{1}{3}}}.$$

Thus,

$$\frac{21}{13} = 1 + \frac{1}{1 + \frac{1}{1 + \frac{1}{1 + \frac{1}{1 + \frac{1}{1 + \frac{1}{1}}}}}}.$$

One method to solve quadratic equations is to use continued fractions.

$$x^2 + x - 6 = 0.$$

We begin with

$$x = -1 + \frac{6}{x}.$$

Then express x in the right hand side by $-1 + \frac{6}{x}$ :

$$x = -1 + \frac{6}{-1 + \frac{6}{x}}.$$

We repeat this substitution again

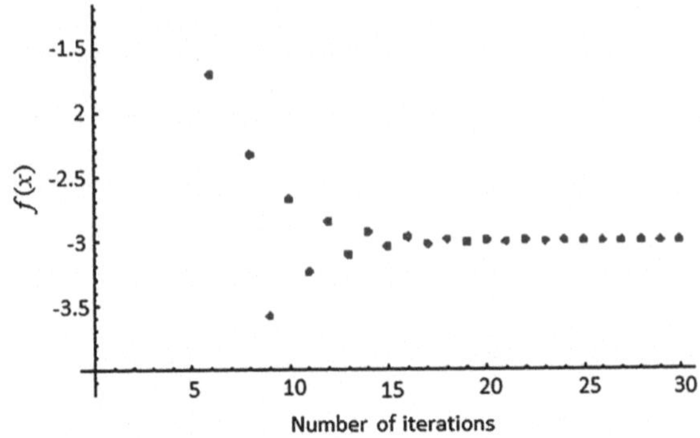


**Fig. 1.8** Approximation of the roots $f(x) = x^2 + x - 6$

$$x = -1 + \frac{6}{-1 + \frac{6}{-1+\frac{6}{x}}}.$$

We now generate an infinitely continuous fraction that approximates almost one of the roots. Approximate one of the roots $f(x) = x^2 + x - 6$ with various iterations is shown in the Fig. 1.8.

## 1.5 Diophantine Equations

**Definition** Suppose $P(x, y, \ldots)$ is a polynomial with several variables such that its coefficients are integers. Diophantine equation is an algebraic equation as follows:

$$P(x, y, z, \ldots) = 0.$$

Consider the following equations:

$$2x + 3y = 11$$
$$7x^2 - 5y^2 + 2x + 4y - 11 = 0$$
$$y^3 + x^3 = z^3.$$

The question that needs to be addressed is whether the Diophantine equation in the domain of integers has an answer.

In 1900, German mathematician David Hilbert raised 23 of the most significant unsolved problems of the twentieth century. His tenth problem was the solvability of a general Diophantine equation. Hilbert proposed a general method for solving all Diophantine equations.

What is the solvable concept? What is the meaning of algorithm?

**1930.** The concept of computation was developed by Kurt Gödel, Stephen Cole Kleene, and Alan Turing.

**1946.** Turing invented the Turing machine and proposed fundamental unsolvable problems.

**1970.** Yuri Matiyasevich showed that Diophantine problem could not be solved.

**heorem** *There is no algorithm for Diophantine's arbitrary equation to determine whether this equation has a solution.*

### *1.5.1 Linear Diophantine Equations*

**Definition** The linear Diophantine equation based on the two variables x and y is an equation as follows:

$$ax + by = c$$

where $a, b, c \in \mathbb{Z}$ are the integer coefficients, which we are looking for the integer solutions.

Obviously, not all of these equations are solvable. For instance

$$2x + 2y = 1$$

there is no integer solution

Some linear Diophantine equations have finite number of solutions, for example

$$2x = 4$$

and some have infinite number of solutions.

**Theorem** *The linear equation $a, b, c \in \mathbb{Z}$*

$$ax + by = c$$

*has an integer solution in x and $y \in \mathbb{Z} \Leftrightarrow$* $\gcd(a, b)|c$.

***Proof*** ($\Rightarrow$) Assume the linear equation $ax + by = c$ for $a, b, c \in \mathbb{Z}$ holds. We show that $\gcd(a, b)|c$. It is clear that

$$\gcd(a, b)|a \wedge \gcd(a, b)|b.$$

Thus,

$$\gcd(a, b)|(xa + yb) \Rightarrow \gcd(a, b)|c.$$

($\Leftarrow$) Given

$$\gcd(a, b)|c \Rightarrow \exists z \in \mathbb{Z}, c = \gcd(a, b) \times z.$$

Multiply this by z:

$$z \times \gcd(a, b) = a \times x_1 \times z + b \times y_1 \times z$$
$$c = a \times x_1 \times z + b \times y_1 \times z.$$

Then the pair $x_1 \times z$ and $y_1 \times z$ is the solution. QED.

#### 1.5.1.1 How Do You Find a Particular Solution?

Consider the following linear Diophantine equation

$$ax + by = c.$$

By extended Euclidean algorithm we find GCD and such $n$ and $m$ that

$$a \times n + b \times m = \gcd(a, b).$$

Multiply this by c

$$a \times n \times c + b \times m \times c = \gcd(a, b) \times c.$$

Divide it by GCD

$$a\frac{n \times c}{\gcd(a, b)} + b\frac{m \times c}{\gcd(a, b)} = c.$$

Compare this with the original equation

$$ax + by = c.$$

It follows that a particular solution is

$$x_0 = \frac{n \times c}{\gcd(a, b)}; y_0 = \frac{m \times c}{\gcd(a, b)}.$$

***Question*** Are $x_0$ and $y_0$ integers?

***Example*** Find a particular solution of

$$56x + 72y = 40.$$

*Solution* Use the extended Euclidean algorithm to find GCD, $n$ and $m$.

$$\gcd(56, 72) = 8 = 4 \times 56 + (-3) \times 72.$$

Then one of the solutions is

$$x_0 = \frac{4 \times 40}{8}; y_0 = \frac{(-3) \times 40}{8}$$
$$x_0 = 20; y_0 = -15.$$

#### 1.5.1.2 How Do You Find All Solutions?

By the extended Euclidean algorithm we find GCD and such $n$ and $m$ that

$$\gcd(a, b) = a \times n + b \times m$$
$$\gcd(a, b) \times c = a \times n \times c + b \times m \times c.$$

Next we add and subtract $a \times b \times k$, where $\forall k \in \mathbb{Z}$

$$\gcd(a, b) \times c = a \times n \times c + b \times m \times c + a \times b \times k - a \times b \times k.$$

Collect terms with respect $a$ and $b$

$$a \times (nc + bk) + b \times (mc - ak) = \gcd(a, b) \times c.$$

Divide this by $\gcd(a, b)$

$$a \times \frac{(nc + bk)}{\gcd(a, b)} + b \times \frac{(mc - ak)}{\gcd(a, b)} = c.$$

It can be rewritten as

$$c = a \times \left(\frac{nc}{\gcd(a, b)} + \frac{bk}{\gcd(a, b)}\right) + b \times \left(\frac{mc}{\gcd(a, b)} - \frac{ak}{\gcd(a, b)}\right)$$

or

$$c = a \times \left(x_0 + \frac{bk}{\gcd(a, b)}\right) + b \times \left(y_0 - \frac{ak}{\gcd(a, b)}\right)$$
$$k = 0, \pm 1, \pm 2, \ldots$$

since $(x_0, y_0)$ is a particular solution.

Therefore, all integer solutions are in the form

$$x = x_0 + \frac{bk}{\gcd(a, b)}, y = y_0 - \frac{ak}{\gcd(a, b)}$$
$$k = 0, \pm 1, \pm 2, ...$$

***Example*** Find all integer solutions of

$$56x + 72y = 40.$$

*Solution* Use the extended Euclidean algorithm to find GCD, $n$ and $m$.

$$\gcd(56, 72) = 8 = 4 \times 56 + (-3) \times 72.$$

All solutions are in the form

$$x = \frac{nc}{\gcd(a, b)} + \frac{bk}{\gcd(a, b)}$$
$$y = \frac{mc}{\gcd(a, b)} - \frac{ak}{\gcd(a, b)}.$$

Hence,

$$x = \frac{4 \times 40}{8} + \frac{72k}{8} = 20 + 9k$$
$$y = \frac{-3 \times 40}{8} - \frac{56k}{8} = -15 - 7k.$$

#### 1.5.1.3 Positive Solutions of LDE

In some applications it may be necessary to find all the positive solutions $x, y \in \mathbb{Z}^+$.

We take a general solution

$$x = \frac{nc}{\gcd(a, b)} + \frac{bk}{\gcd(a, b)}, k \in \mathbb{Z},$$
$$y = \frac{mc}{\gcd(a, b)} - \frac{ak}{\gcd(a, b)}, k \in \mathbb{Z},$$

from which we get two inequalities

$$nc + bk > 0$$
$$mc - ak > 0.$$

To understand the equation of several positive solutions, we consider two cases.

1. $ax + by = c$, $\gcd(a, b) = 1, a, b > 0$.
2. $ax - by = c$, $\gcd(a, b) = 1, a, b > 0$.

Based on the first case, the equation has a finite solution

$$-\frac{nc}{|b|} < k < \frac{mc}{|a|}.$$

In the latter case, there are infinite solutions

$$nc - |b|k > 0$$
$$mc - |a|k > 0.$$

***Example*** Find the number of positive integers solutions

$$4x + 7y = 117$$
$$\gcd(4, 7) = 1 = 2 \times 4 + (-1) \times 7.$$

The number of positive integers solutions be found from the following system

$$nc + bk > 0$$
$$mc - ak > 0$$

and it is converted as follows for the desired equation:

$$2 \times 117 + 7 \times k > 0$$
$$(-1) \times 117 - 4 \times k > 0.$$

Therefore,

$$-\frac{2 \times 117}{7} < k < \frac{-117}{4}.$$

There 4 such $k$, namely $k = -33, -32, -31, -30$

#### 1.5.1.4 LDEs with Three Variables

Consider the equation

$$3x + 6y + 5z = 7.$$

After factorizing

$$\gcd(3, 6)(x + 2y) + 5z = 7.$$

Let

$$w = x + 2y.$$

The equation becomes.

$$3w + 5z = 7$$

Its general solution is

$$w = 2 \times 7 + 5k$$
$$z = (-1) \times 7 - 3k$$

since

$$\gcd(3, 5) = 1 = 2 \times 3 + (-1) \times 5.$$

Next we find $x$ and $y$

$$x + 2y = 14 + 5k.$$

Since $\gcd(1, 2) | (14 + 5k)$, the equation is solvable and the solution is

$$x = 1 \times (14 + 5k) + 2 \times l$$
$$y = 0 \times (14 + 5k) - 1 \times l$$

where $l \in \mathbb{Z}$ is another parameter. Here are all triple-solutions

$$x = 5k + 2l + 14$$
$$y = -l$$
$$z = -7 - 3k$$

where $k, l = 0, \pm 1, \pm 2, ...$

#### 1.5.1.5 LDEs with Several Variables

Consider the equation

$$3x + 4y + 5z + 6t = 7.$$

Let

$$u = 3x + 4y \text{ and } v = 5z + 6t.$$

Solve

$$u + y = 7.$$

We find

$$u = 7 + k_1 \text{ and } v = -k_1.$$

Next, we solve

$$\begin{aligned} 3x + 4y &= 7 + k_1 \\ 5z + 6t &= -k_1. \end{aligned}$$

The solution will have two new parameters $k_2$ and $k_3$

$$\begin{aligned} x &= -(7 + k_1) + 4k_2 \\ y &= (7 + k_1) - 3k_2 \\ y &= k_1 + 6k_3 \\ t &= -k_1 - 5k_3. \end{aligned}$$

where $\forall k_1, k_2, k_3 \in \mathbb{Z}$.

#### 1.5.1.6 The Money-Changing Problem

**Problem** How many ways are there to make change for a **$1** bill? In the change-making problem, there given a finite set of denominations and unlimited supply of coins. We need to represent a given value with the fewest number of coins. Here are all possible choices to make change for a dime

$$\begin{gathered} 1+1+1+1+1+1+1+1+1+1 \\ 1+1+1+1+1+5 \\ 5+5 \\ 10. \end{gathered}$$

The above problem is a simple form of the linear Diophantine problems. For a given integer $N$ find all solutions in non-negative integers to the equation

$$\begin{aligned} &x_1 + 5x_2 + 10x_3 + 25x_4 + 50x_5 + 100x_6 = N \\ &\min_{x_k \geq 0} \sum_{k=1}^{6} x_k. \end{aligned}$$

Another approach is to use a recursive function.
Split the problem into two sub-problems.

1. Compute the number of changes with the largest coin.
2. Compute the number of changes without the largest coin.

#### 1.5.1.7 The Greedy Algorithm

Repeatedly take the largest coin that will fit and recursively give change for the remainder.

The algorithm gives you the smallest number of coins possible.

Let us imagine a NEW coin system. It uses 4, 7, 9 and 34-cent coins. The system has an obvious flaw that makes our system more interesting. Not all amounts can be exchanged. Which amounts can be exchanged? What is the largest amount that cannot be exchanged?

Given $a, b \in \mathbb{N}^+$. Which $z \in \mathbb{N}^+$ can be expressed as a linear combination of $a$ and $b$

$$z = ax + by$$

with positive integers $x$ and $y$.

#### 1.5.1.8 The Frobenius Coin-Exchange Problem.

Given coins of k different denominations

$$0 < d_1 < d_2 < \cdots < d_k and \ \gcd(d_1, ..., d_k) = 1.$$

Assuming an infinite supply of coins, we want to know what is the largest amount of

$$d_1x_1 + d_2x_2 + \cdots + d_kx_k = N$$

that cannot be exchanged. This number is called the Frobenius number.

We will derive a formula for Frobenius numbers in case of two denominations. Consider

$$z = ax + by, \gcd(a, b) = 1.$$

We need to find the largest $z$ such that the equation has NO positive solutions in $x$ and $y$.

When the Diophantine equation has a unique positive solution?

$$x = x_0 + b \times k$$
$$y = y_0 - a \times k$$
$$k = 0, \pm 1, \pm 2, ...$$

Among all x-solutions

$$..., x_0 - b, x_0, x_0 + b, ...$$

we choose one that

$$x_0 \leq x < b.$$

What about y-solutions?

$$y \geq 0.$$

These are necessary conditions!

***Example*** Consider

$$3x + 5y = 11.$$

All solutions are

$$x = \frac{nc + bk}{\gcd(a, b)}, y = \frac{mc - ak}{\gcd(a, b)}$$

or

$$x = 22k + 5k, y = -11 - 3k.$$

Among all $x$-solutions we choose such that

$$0 < x < b.$$

Therefore,

$$0 \leq 22 + 5k < 5 \Rightarrow k = -4,$$

and the solution is $x = 2$, $y = 1$.

**heorem** *Let a and b be relatively prime positive integers. That the largest positive integer that is not representable in the form a $ax + by$ is defined* $(a - 1)(b - 1) - 1$.

***Proof*** Consider

$$U = \{ax + by | 0 \le x < b \wedge y \ge 0\}$$
$$V = \{ax + by | 0 \le x < b \wedge y < 0\}.$$

If $z \in U \Rightarrow z$ can be represented in the form $z = ax + by$.
If $z \in V \Rightarrow z$ is not representable in the form $z = ax + by$.
What is the maximum value in set $V$, $\max(V)$?

$$x = b - 1 \wedge y = -1.$$

It follows that the maximum $z$ that is not representable is

$$z = (b - 1)a + (-1)b.$$

This is the Frobenius number. QED.

The problem is algorithmically solved for 3 denominations and completely open for 4 and more denominations. The problem is NP-hard (Ramirez-Alfosin 1996).

#### 1.5.1.9 The Frobenius Postage Stamp Problem

The postal service of the rebuilt Iraq wants to issue 5 denominations of stamps (in whole numbers of dinars) so that by using no more than 3 stamps on an envelope, citizens can pay for any postage from 1 to 36 dinars. How can this be done?

The set of stamps is 1, 4, 6, 14, 15.

$$\begin{aligned}
1 &= 1 \\
2 &= 1 + 1 \\
3 &= 1 + 1 + 1 \\
&\vdots \\
19 &= 14 + 4 + 1 \\
&\vdots \\
36 &= 15 + 15 + 6 \\
37 &= ?
\end{aligned}$$

The set of stamps is 1, 4, 6, 14, 15.

We like to generalize problems. Given stamps of $k$ different denominations $1 = d_1 < d_2 < \cdots < d_k$.

We want to know what is the smallest integer $N$ that cannot be represented by

$$d_1x_1 + d_2x_2 + \cdots + d_kx_k = N$$
$$x_1 + x_2 + \cdots + x_k \le H$$

where $H \geq a_2 - 2$ is the number of stamps. The two stamps problem can be solved

$$(H + 3 - d_2)d_2 - 1.$$

In the general case, it is NP-hard (Shalit 2002).

## 1.6 Binomial Expansion

There are different methods for determining binomial coefficients. We select one that describes the name. In fact, a binomial is an algebraic term that include two expressions, for instance, $x + y$. We are going to multiply binomials

$$(x + y)^2 = (x + y)(x + y) = 1x^2 + 2xy + 1y^2$$
$$(x + y)^3 = (x + y)^2(x + y) = 1x^3 + 3x^2y + 3xy^2 + 1y^3$$
$$(x + y)^4 = (x + y)^3(x + y) = 1x^4 + 4x^3y + 6x^2y^2 + 4xy^3 + 1y^4.$$

Numbers that appear as coefficients of binomial extension are called binomial coefficients. The binomial coefficients $n$ and $k$ are also written.

$$C(n, k) or \binom{n}{k}$$

and read as "$n$ choose $k$". To describe the latter name consider the following quadratic equation

$$(x + y)^2 = (x + y)(x + y) = \binom{2}{0}x^2y^0 + \binom{2}{1}xy + \binom{2}{0}y^2x^0.$$

Consider each two terms in $(x + y)(x + y)$ as a sentence containing two variables, $x$ and $y$. To form the expansion, we need to select two $x$'s from both boxes, and no $y$ is required. We have only one choice to do this. When forming the expression $xy$ we have two choices: either $x$ from the first box and $y$ from the second or vice versa.

$$(x + y)^3 = (x + y)(x + y)(x + y) = \binom{3}{0}x^3 + \binom{3}{1}x^2y + \binom{3}{1}xy^2 + \binom{3}{0}y^3.$$

In the case of the cubic form $(x + y)^3$ we have three boxes. The coefficient in $x^2y$ means the number of options to choose two $x$ and one $y$. All we have to do is decide which box to select $y$ from and then select the $x$ from all the other boxes. Therefore, from a set of 3 boxes the coefficient $x^2y$ is the number of ways we can select one $y$.

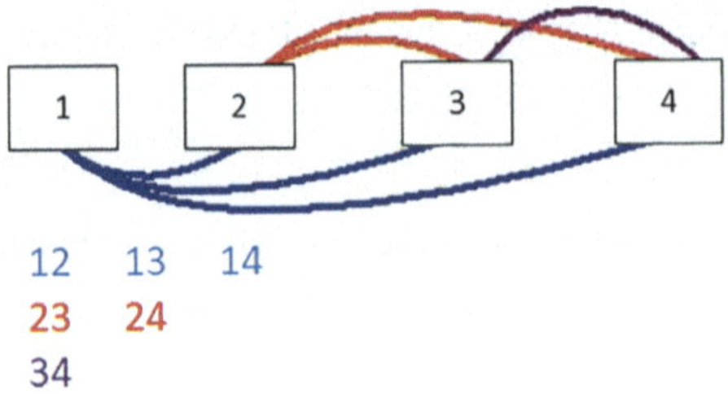


**Fig. 1.9** The coefficient of $x^2y^2$

$$(x+y)^4 = \binom{4}{0}x^4 + \binom{4}{1}x^3y + \binom{4}{2}x^2y^2 + \binom{4}{1}xy^3\binom{4}{0}y^4.$$

We have four boxes for $(x + y)^4$. The coefficient $x^3y$ means the number of choices to pick one $y$ from four boxes. There are three choices to do so. The coefficient by $x^2y^2$ means the number of choices two pick from 4 boxes. This number is 6 as it follows from the Fig. 1.9.

### *1.6.1 Combinatorics*

Binomial coefficients are important in combinations that provide formulas for certain counting problems. In this section we get a formula for calculating $C(n, k)$.

We want to start with the ordered sets. On the other hand, in an ordered set, there is a first element, a second element, etc. Here, the ordered sets are arranged as two elements $x$ and $y$:

$$\{x, y\}, \{y, x\}.$$

Consider the ordered sets with x, y and z. In this case, we have six sets as follows:

$$\{x, y, z\}, \{x, z, y\}, \{y, x, z\}, \{z, x, y\}, \{y, z, x\}, \{z, y, x\}.$$

How many ordered sets with $n$ elements can be made?

In general, the number of ordered sets with $n$ elements is equal to $n!$. We can show this as follows.

The number of ways to put an element into the first situation is n. When the first situation is filled, there is an $n - 1$ option to select one element from the rest for the second situation, and so on. This number is called the arrangements number. The decision tree below simplifies the visualization of the proof (Fig. 1.10).

Every path from root to leaf represents an ordered set.

In the following, we count the number of methods for obtaining an ordered subset with $k \leq n$ elements from a set with $n$ elements. This is the number of permutations indicated by $P(n, k)$. This reads the of permutations number of $n$ taken by $k$. Drawing

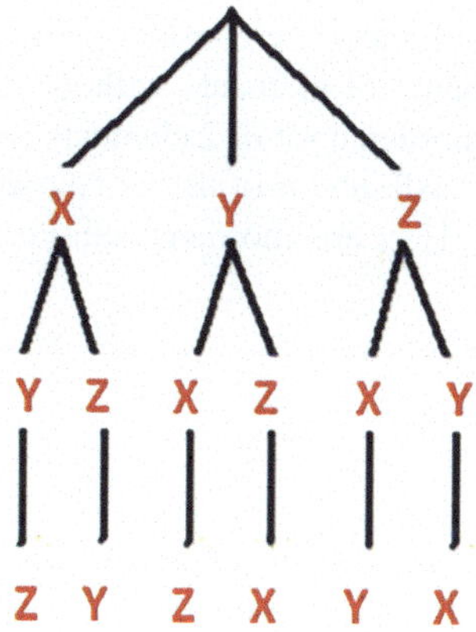


**Fig. 1.10** A diagram of arrangements number

the decision tree, we visualize this number. Here are some subsets with size 2 which can be made from a set with length 4 (Fig. 1.11).

Every path is arranged from root to leaf corresponding to the ordered set. Therefore, if $P(4, 2) = 12$, the number of ordered sets of size 3 is made up of a set of size 4, then another level of nodes is drawn (Fig. 1.12).

Thus, $P(4, 3) = 24$. In general,

$$P(n, k) = n(n-1)...(n-k+1) = \frac{n!}{(n-k)!}.$$

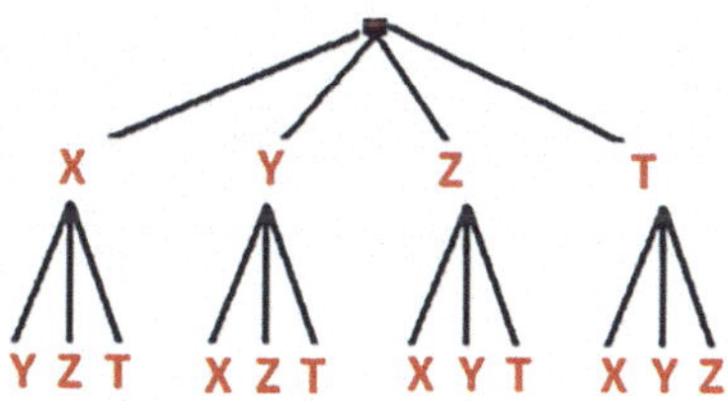


**Fig. 1.11** A diagram of permutations number

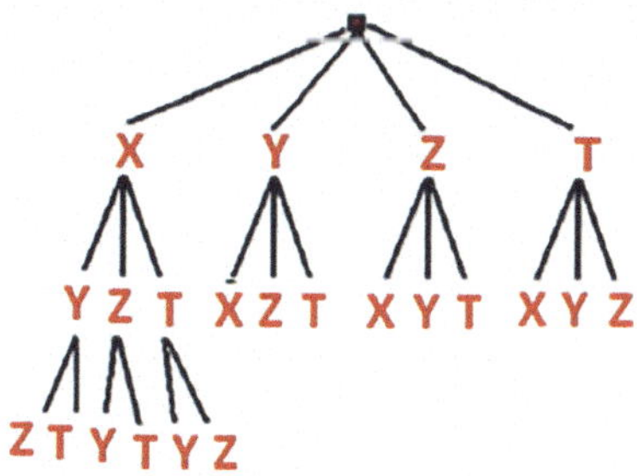


**Fig. 1.12** A generalized diagram of permutations number

Finally, we consider permutations of unordered sets. Combinations are arrangements of elements without regard to their order or position. There is only one unordered set of 2 elements $\{x, y\}.\{x, y\}$.

All subsets of size $k$ of the set of $n$ elements is denoted by $C(n, k)$, such that $C(n, n) = 1$. Here are unordered subsets of 2 elements out of $\{x, y, z, t\}$ :

$$\{x, y\}, \{x, z\}, \{x, t\}, \{y, z\}, \{y, t\}, \{z, t\}$$

$$C(4, 2) = 6.$$

In order to obtain the formula $C(n, k)$, we reconsider finding the number of ordered sets with size $k$ in a set of size $n$. We can solve this problem in another way. This is the algorithm:

- Create all unordered subsets of size $k$.
- Order the elements of these subsets.

Consider all unordered subsets of 2 elements out of $\{x, y, z, t\}$ :

$$\{x, y\}, \{x, z\}, \{x, t\}, \{y, z\}, \{y, t\}, \{z, t\}.$$

How can we order elements in the any subset? The considered order is simple. We directly order the elements and next we do in reverse ways:

$$\begin{aligned}
&\{x, y\}, \{x, z\}, \{x, t\}, \{y, z\}, \{y, t\}, \{z, t\} \\
&\{y, x\}, \{z, x\}, \{t, x\}, \{z, y\}, \{t, y\}, \{t, z\} \\
&C(2, 4) \times 2 = P(2, 4).
\end{aligned}$$

We now consider all subsets of 3 elements out of $\{x, y, z, t\}$ : $\{x, y, z, t\}$ :

$$\{x, y, z\}, \{x, z, t\}, \{y, z, t\}, \{x, z, t\}.$$

There are six choices to order each subset. Here are choices for the first subset:

$$\{x, y, z\}, \{x, z, y\}, \{y, x, z\}, \{z, x, y\}, \{y, z, x\}, \{z, y, x\}$$

$$C(4, 3) \times 6 = P(4, 3).$$

Therefore, we demonstrated that

$$C(n, k)P(k, k) = P(n, k).$$

It follows that

$$C(n,k) = \frac{P(n,k)}{P(k,k)} = \frac{n!}{(n-k)!k!}.$$

***Example*** (the game of poker). How many complete different houses are there in poker? There are 13 sects: King and Ace. There are 4 suits: shovel, heart, diamond, and stick. The complete house contains 3 units + 2 other sects.

*Solution* Choose a name for 3 sets. Then choose a name for the 2 sets. These can be done in $13 \times 12$ methods. Then, we have to choose 3 cards out of 4 and 2 cards out of 4. Also, the total number of different houses is equal to:

$$13 \times 12 \times C(4,3) \times C(4,2) = 3744.$$

**Theorem**

$$C(n,k) = C(n, n-k).$$

***Proof***

$$C(n,k) = \frac{n!}{(n-k)!k!}$$

$$C(n, n-k) = \frac{n!}{(n-(n-k))!(n-k)!} = \frac{n!}{(n-k)!k!}.$$

From the binomial expansions, we have.

$$(x+y)^3 = \binom{3}{0}x^3 + \binom{3}{1}x^2y + \binom{3}{1}xy^2 + \binom{3}{0}y^3.$$

On the other hand,

$$\binom{3}{1} = \binom{3}{2}, \binom{3}{0} = \binom{3}{3},$$

therefore,

$$(x+y)^3 = \binom{3}{0}x^3 + \binom{3}{1}x^2y + \binom{3}{2}xy^2 + \binom{3}{3}y^3.$$

## 1.7 Modular Arithmetic

**Definition** *A* mod *n* means the remainder when *a* is divided by *n*.

$$a = qn + r.$$

**Definition (modulo equivalence).**

$$a \equiv b \bmod n \textit{ if and only if } n|(a-b).$$

Let say that a and b are equivalent modulo n. We will also write modulo equivalence as

$$a \equiv_n b.$$

**Theorem** *The symbol $\equiv_n$ is an equivalence relation on the integers.*

*An equivalence class consists of those integers which have the same remainder on division by n. The equivalence classes are also known as congruence classes modulo n. Rather than say the integers a and b are equivalent, we say that they are congruent modulo n.*

**Definition** The set of all integers congruent to $a$ modulo $n$ is called the residue class $[a]$.

***Example*** Residue classes mod 3:

$$[0] = \{..., -6, -3, 0, 3, 6, ...\}$$
$$[1] = \{..., -5, -2, 1, 4, 7, ...\}$$
$$[2] = \{..., -4, -1, 2, 5, 8, ...\}$$

**Definiton** The set of residue classes is denoted by *Zn*.

$$Z_n = \{0, 1, 2, ..., n-1\}.$$

**Theorem** (arithmetic on Zn). *When we do +, − or modulo n, we can swap another number in the remaining class with a number.*

***Example*** Calculate.

$$414 \times 463 \bmod 413$$
$$1 \times 50 = 50.$$

Note that although the arithmetic in *Zn* seems to be the same as *Z*, multiplying two nonzero elements of *Zn* can somewhen be zero (cancellation feature).

***Question*** Are the following statements correct?

$$xa \equiv_n xb \Rightarrow a \equiv_n b.$$

Only if $\gcd(x, n) = 1$.

***Proof*** $Xa \equiv_n xb \Rightarrow n|(xa - xb) \Rightarrow n|x(a-b) \Rightarrow n|(a-b) \Rightarrow a \equiv_n b.$

**Definition** $Z_n^* = \{x \in Z_n | \gcd(x, n) = 1\}$.

***Example***

$$Z_0 = \{0, 1, 2, 3, 4, 5\}$$
$$Z_6^* = \{1, 5\}.$$

The advantage of $Z_n^*$ is that it has a cancellation feature.

**Definition** Euler's phi function (or totient function) is the size of $Z_n^*$

$$\phi(n) = |Z_n^*|,$$

where $\phi(n)$ is the number of integers $1 \leq k \leq n$ coprime to $n$.

***Example***

$$\phi(12) = |\{1, 5, 7, 11\}| = 4.$$

It is quickly inferred that for every prime number $p$, we have $\phi(p) = p - 1$.

*Note* Multiplicative inverse of a number, i.e. $x^{-1}$ is expressed as follows:

$$x \times x^{-1} = 1$$

is another difference between *Zn* and Z. *Z* has only two inverse elements: 1 and − 1. We have proved that if *p* is prime, every element has an inverse in *Zp*. We express that *Zp* is a finite field. A set in which we can apply $+, -, \times, /$. The following theorem addresses this issue.

**heorem** (Fermat's little theorem). *For every prime number p, which p* $\not| a$*, we have:*

$$a^{p-1} \equiv_p 1.$$

*Assuming p is prime and based on this theorem, the inverse is expressed by*

$$a^{-1} - a^{p-2}.$$

## 1.7.1 Applications of Modular Arithmetic

**Problem 1** How do we efficiently store people's records?

If we use Social Security number as the key, we will have to deal with an array of size $10^{10}$. $10^{10}$.

The solution: hashing. Store records in the table at index $h(k)$ defined by

$$h(k) = k(\text{mod}N).$$

Here, $h$ is a hash function, and $N$ is an array size. Regardless of chosen $N$, collisions might happen

$$h(k_1) = h(k_2).$$

**Problem** How do we generate a random number?

Suppose random () or rand () are functions about which there is nothing random. Generating a quasi-random sequence can be used as a standard technique. The linear congruential method is the most common method used to generate quasi-random numbers. We generate a sequence of pseudo-random numbers $x_k$ using the following sequence

$$x_{n+1} = ax_n + b(\text{mod}m), n > 0$$

$x_0 < m$ is the seed.

with suitable values of $0 < a < m, 0 \leq b < m, m > 0$ and $\gcd(b, m) = 1$.

For instance,

$$\begin{aligned} x_{n+1} &= 7x_n + 4(\text{mod}9), n > 0, \\ x_0 &= 3. \end{aligned}$$

This generates the following sequence

$$\begin{aligned} x_1 &= (7x_0 + 4) = (7 \times 3 + 4) = 25 = 7(\text{mod}9), \\ x_2 &= ((7x_1 + 4) = (7 \times 7 + 4) = 53 = 8(\text{mod}9), \\ x_3 &= (7x_2 + 4) = (7 \times 8 + 4) = 60 = 6(\text{mod}9), \end{aligned}$$

and so on.

This recurrence is ultimately periodic. Therefore, the use of recurrences with a long period is desirable.

The information is decoded in the first 9 digits. The last digit is for parity check

$$a_1 + 2a_2 + \cdots + 9a_9 = a_{10}(\text{mod}11).$$

Applying this to our textbook

$$\begin{aligned} &1 \times 0 + 2 \times 1 + 3 \times 3 + 4 \times 1 + 5 \times 8 + 6 \times 4 + 7 \times 8 + 9 \times 6 + 9 \times 8 = 255 \\ &= 2(\text{mod}11). \end{aligned}$$

The total of 9 digits is 255 and it equals to the last digit mod 11.

**Problem** The Gauss Easter formula.

As a moveable feast, the date of Easter is determined in each year through a calculation known as computus. Easter is celebrated on the first Sunday after the Paschal full moon, which is the first full moon on or after 21 March. Carl Friedrich Gauss provided a precise formula for calculating the Easter date, which is summarized as follows:

$a = year(\bmod 11)$

$b = year(\bmod 4)$

$c = year(\bmod 7)$ where $M = 24, N = 5$ are constants. In this case,

$d = 19a + M\,(\bmod 30)$

$e = 2b + 4c + 6d + N\,(\bmod 7)$

the formula is valid throughout the period between 1900 and 2099 years. Then, after 2099, Easter falls on $(22 + d + e)$ March or $(d + e - 9)$ April.

In 2006 the Catholic Church celebrated Easter on April 16. Dates are according the Gregorian calendar.

### *1.7.2 Solving Linear Congruences*

**efinition** A relation of the form

$$a \times x \equiv b(\bmod n)$$

is called a linear congruence. We will also consider this as

$$a \times x \equiv_n b.$$

How many solutions are there? Since each element of a congruent class is also a solution, $h(k)$ is a solution.

***Example***

$$x \equiv 3(\bmod 4)\cdot$$

Here are solutions:

$$\ldots, -1, 3, 7, 10, \ldots$$

Thus, uniqueness can only be mod n.

$$x = 3 + 4k, \text{ where } k \in \mathbb{Z}.$$

***Example***

$$2x \equiv 2(\bmod 4).$$

The congruence is satisfied for two representatives.

$$x = 1 \, and \, x = 3.$$

Therefore, the equation has two solutions

$$x = 1 + 4k \text{ and } x = 3 + 4k, \text{ where } k \in \mathbb{Z}.$$

**heorem** The linear congruence.

$$a \times x \equiv b(\text{mod} n)$$

has a unique solution if $\gcd(a, n) = 1$.

***Proof*** ($\Leftarrow$) First we prove that the equation has a solution. To solve the equation means to find the inverse of a

$x \equiv b \times a^{-1}(\text{mod} n)$.

But, as we know, the inverse does not necessarily exist in *Zn*. Therefore, some restrictions are required. In the next paragraph we show how to find an inverse using the extended Euclidean algorithm.

Given $\gcd(a, n) = 1$, it follows by the extended Euclidean algorithm that $\exists s, r \in \mathbb{Z}$

$$a \times r + n \times s = 1.$$

Consider this equation mod n.

$$a \times r \equiv 1(\text{mod} n)$$

This means that $r$ is the multiplicative inverse of a.

$$a \equiv r^{-1}(\text{mod} n)$$

Therefore, the solution is

$$x \equiv b \times r(\text{mod} n).$$

Next, we prove uniqueness. We assume that there are two solutions, $X$ and $Y$.

$$\left.\begin{aligned} a \times x &\equiv b(\text{mod} n) \\ a \times Y &\equiv b(\text{mod} n) \end{aligned}\right\} \Rightarrow a \times x \equiv a \times Y(\text{mod} n)$$

$$\Rightarrow n|a(X - Y) \Rightarrow n|(X - Y) \Rightarrow X \equiv Y(\text{mod} n).$$

The proof in ($\Rightarrow$) direction is left as an exercise.

### 1.7.3 Chinese Remainder Theorem

Consider the linear congruences as follows:

$$\begin{cases} x \equiv 2(\bmod 3) \\ x \equiv 3(\bmod 5) \\ x \equiv 2(\bmod 7) \end{cases}$$

Does a solution exist? Is it unique?
Let us build a solution by solving the equation $x \equiv 2(\bmod 3)$ at first step.

$$y_1 \equiv 2(\bmod 3).$$

Assuming that the solution is 0 modulo 5 and 7

$$y_1 \equiv 2(\bmod 5)$$
$$y_1 \equiv 0(\bmod 7).$$

The straightforward answer is defined by

$$5 \times 7 \times z \equiv 2(\bmod 3).$$

Therefore,

$$z - 1 \Rightarrow y_1 = 35.$$

Then, assuming that the answer is 0 for modules 3 and 7, we again solve the second equation $x \equiv 3(\bmod 5)$

$$3 \times 5 \times z \equiv 3(\bmod 5).$$

We have

$$z = 3 \Rightarrow y_2 = 63.$$

Likewise, solving the last equation gives $x \equiv 2(\bmod 7)$

$$3 \times 5 \times z \equiv 2(\bmod 7)$$
$$z = 2 \Rightarrow y_3 = 30.$$

Therefore, the general solution is

$$y_1 + y_2 + y_3 = 35 + 63 + 30 = 128.$$

Obviously, this is the solution of $3 \times 5 \times 73 \times 5 \times 7$

$$x \equiv 128(\bmod 105).$$

In the parametric form

$$x = 23 + 105 \times k, k \in \mathbb{Z}.$$

**heorem** (Chinese Remainder Theorem). *Let $m_1, m_2, ..., m_n \in \mathbb{N}^+$ be pairwise relatively prime. The system*

$$x = a_k(\bmod m_k), k = 1, 2, ..., n$$

*has a unique solution modulo $M = m_1 \times m_2 \times ..., m_n$.*

***Proof*** First we prove that the system has a solution. Proceeding as in the above example, we define solution to each equation as

$$y_k \equiv a_k(\bmod m_k)$$
$$y_k \equiv 0\big(\bmod m_j\big), k \neq j.$$

Combining them together yields the equation.

$$\frac{M}{m_k} z \equiv a_k(\bmod m_k)$$

Given z, there is definitely a solution (why?). Therefore,

$$y_k \equiv \frac{M}{m_k} z.$$

We now define the general solution as follows:

$$y_1 + y_2 + \cdots + y_n.$$

Obviously, this is a solution for a system determined by modulus M.

**Uniqueness** Assume there are two solutions such as $X$ and $Y$. In this case we have $X \equiv_{m_k} Y$ for all $k$.
Thus, they have identical M and we can write

$$X \equiv_M Y.$$

### 1.7.4 Arithmetic with Large Integers

**Problem** Calculate 7! without facing a number greater than $2^8$.

For this purpose, we first gain a relatively prime module $m_k$ such that $m_1 m_2 ... m_n >$ 7!

We consider 13, 11, 9, and 7.

Then, we do all the calculations, applying these modules

$$\begin{aligned}
&2 \times 5 = (10, 10, 1, 3) \\
&3 \times 4 = \{12, 1, 3, 5\} \\
&5! = \{120, 10, 3, 15\} = \{3, 10, 3, 1\} \\
&6! = \{18, 60, 18, 6\} = \{5, 5, 0, 6\} \\
&7! = \{35, 35, 0, 42\} = \{9, 2, 0, 0\}.
\end{aligned}$$

Using CRT, we retrieve 7! in the last step as follows:

$$\begin{cases} x \equiv 9 (\bmod 13) \\ x \equiv 2 (\bmod 11) \\ x \equiv 0 (\bmod 9) \\ x \equiv 0 (\bmod 7). \end{cases}$$

We find the number x such that $0 < x < 13 \times 11 \times 9 \times 7$.

*Note* Based on the system above, numbers greater than $2^8$ are not be generated.

Here is another way to solve the congruences system. The list of modules and residues is as follows:

$$\begin{aligned}
m &= \{13, 11, 9, 7\} \\
r &= \{9, 2, 0, 0\}
\end{aligned}$$

respectively. In the first step, the inverses of each module relative to each subsequent module in the list is considered. This can be investigated with the extended Euclidean algorithm as follows:

$$\begin{aligned}
&13^{-1} \equiv 6 \bmod 11, 13^{-1} \equiv 7 \bmod 9, 13^{-1} \equiv 6 \bmod 7 \\
&11^{-1} \equiv 5 \bmod 9, 11^{-1} \equiv 2 \bmod 7 \\
&9^{-1} \equiv 4 \bmod 7.
\end{aligned}$$

Let us denote the list of inverse (padded with zeros on the left) by

$$inv = \{\{0, 6, 7, 6\}, \{0, 0, 5, 2\}, \{0, 0, 0, 4\}\}.$$

Then,

$$inv_{jk} \times m_j \equiv 1(\bmod m_k).$$

Finally, the solution is built up by

$$x = \sum_{k=1}^{R} rem_k \prod_{j=1}^{k-1} m_j,$$

where $rem = \{9, 2, 0, 0\}$. One can check that $0 \leq x < M = m_1 m_2 ... m_n$. In case of four moduli the solution is

$$x \equiv m_1 + m_1 v_1 + m_1 m_2 v_2 + m_1 m_2 m_3 v_3 (\bmod M)$$

where

$$v_1 = (r_2 - r_1) inv_{1,2} (\bmod m_2)$$
$$v_2 = (r_2 - r_1) inv_{1,2} (\bmod m_2)$$
$$v_3 = \big(\big((r_4 - r_1) inv_{1,4} - v_1\big) inv_{2,4} - v_2\big) inv_{3,4} (\bmod m_2).$$

To investigate that this is a solution, we calculate $x$ mod $m_k, k = 1, 2, 3, 4$. Clearly,

$$a_n = a_{n-1} + (n-1) + \cdots + 2 + 1$$
$$a_n = \frac{n(n+1)}{2}.$$

Due to the fact that every term $x$ except the first term is divisible by $m_1$. We now calculate $x(\bmod m_2)$. Using $\bmod m_2$, we have

$$x \equiv r_1 + m_1 v_1 (\bmod m_2).$$

Substituting $v_1 v_1$

$$x \equiv r_1 + (r_2 - r_1) m_1 inv_{1,2} (\bmod m_2)$$

and applying

$$m_1 \times inv_{1,2} \equiv 1(\bmod m_2),$$

we obtain

$$x \equiv r_2 (\bmod m_2).$$

Generally, choosing the following five relatively prime numbers as moduli

$$\{2^{35} - 1, 2^{34} - 1, 2^{35} - 1, 2^{33} - 1, 2^{29} - 2^{23} - 1\}$$

we can add and multiply positive integers up to the size of $2^{154}$

$$M = \prod_{k=1}^{5} m_k > 2^{154}.$$

## 1.7.5 *The Pigeonhole Principle*

### 1.7.5.1 The Principle

If $m$ pigeons are in $n$ holes and $m > n$, then at least 2 pigeons are in the same hole. In fact, at least $\lceil \frac{m}{n} \rceil$ pigeons must be in the same hole.

- Two or more people in a college have the same birthday.
- In a room with six people in it, there will always be either 3 people who know each other or 3 people who don't know each other.

At any given time in New York City there live at least two people who have the same amount of hairs on their head. Here there are $n = 11$ pigeons in $m = 12$ holes. Since 12 is greater than 11, the pigeonhole principle says that at least one hole has more than one pigeon (Fig. 1.13).

**Fig. 1.13** The pigeonhole principle with 11 pigeons and 12 holes

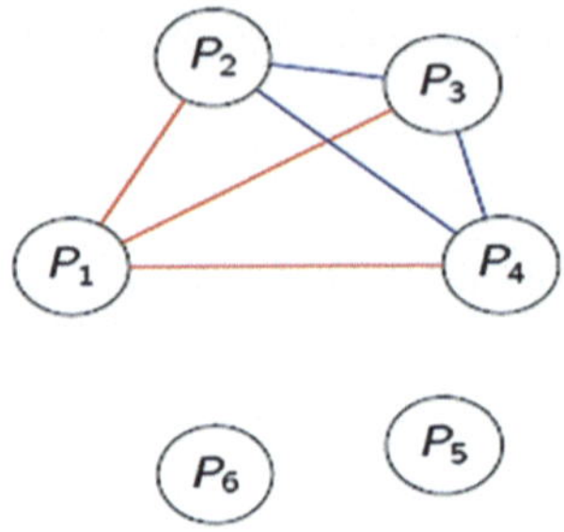


**Fig. 1.14** Graph representation of Trivial problem

***Question*** **(Trivial Applications)**. Show that there are three mutual friends or three strangers in each group of six. (Suppose that two people are either mutual strangers or mutual friends.)

***Proof*** Represent the people as nodes on a graph, and denote friendships using red edges and "stranger-ship" using blue edges. We have to show that there exists a monochromatic triangle. Consider the relationship of $P_1$ to the 5 others. By the pigeonhole principle, 3 of the others must have the same relationship to person 1. Without loss of generality, say $P_2$, $P_3$, $P_4$ are connected to $P_1$ by red edges. Consider the edges between $P_2$, $P_3$, and $P_4$. If any of them are red, then we have a red triangle. Otherwise we have a blue triangle. The graph of six people such that there are either three mutual friends or three strangers is presented in Fig. 1.14.

***Question*** **(Geometric Example)**. Consider any five points $P_1$, ..., $P_5$ in the interior of a square S of side length 1. Show that one can find two of the points at distance at most $\frac{\sqrt{2}}{2}$ apart.

***Proof*** Consider partitioning the square into 4 subsquares as shown. By the pigeonhole principle, 2 of the points must be in one "sub-box." The distance between those two points must be less than the diameter of the sub-box: $\frac{\sqrt{2}}{2}$ (the length of the sub-box's diagonal). This proves the desired result. Five points in the interior of a square S of side length 1 can be represented as follows (Fig. 1.15).

***Question*** **(More Careful "Hole" Construction)**. Suppose that 5 points lie on a sphere. Prove that there exists a closed semi-sphere (half a sphere including boundary), which contains 4 of the points.

***Proof*** Consider the great circle through any two of the points. This partitions the sphere into two hemispheres. By the pigeonhole principle, 2 of the remaining 3 points must lie in one of the hemispheres. These two points, along with the original two points, lie in a closed semi-sphere. Five points on a sphere, such that there is a closed semi-sphere, which contains 4 of the points are graphed in Fig. 1.16.

***Question*** **(Clever Construction of Pigeons and Holes).** A chess master has 77 days to prepare for a tournament. He wants to play at least one game per day, but not more

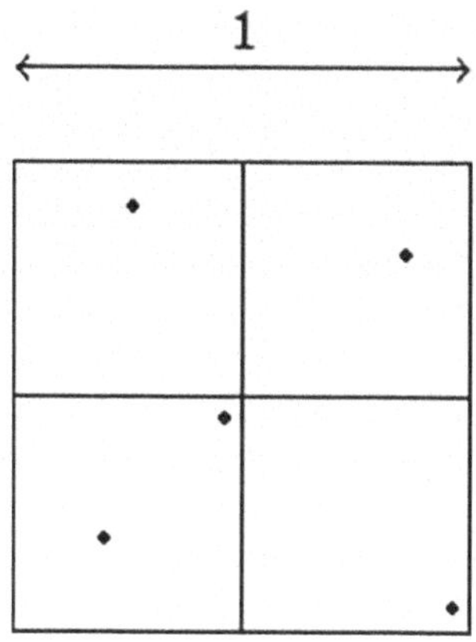


**Fig. 1.15** A representation of five points in the interior of a square with length 1

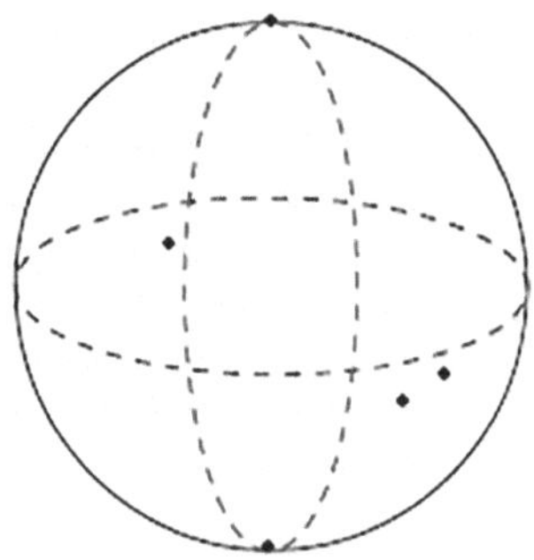

**Fig. 1.16** Geometric representation of More Careful Hole Construction problem

than 132 games. Prove that there is a sequence of successive days on which he plays exactly 21 games.

***Proof*** Dene $S_i$ $(1 \leq i \leq 77)$ as the total number games the chess master plays from day 1 up to day i. Because she plays at least one game a day, $1 \leq S_1 < S_2 < \cdots < S_{77} \leq 132$ (i.e. the $S_i$'s are distinct). Define $T_i = S_i + 21$. Note that the $T_i$'s are all distinct. Now, out of the $S_i$'s and Ti 's, there are $77 \times 2 = 154$ numbers, but these numbers can take at most $132 + 21 = 153$ possible values. By the pigeonhole principle, two of the numbers are equal. This implies that for some i, j, $S_j = T_j = S_i + 21$. Hence, the chess master plays exactly 21 games in the consecutive block from day $i + 1$ to day $j$.

***Question*** **(One Commonly Used Idea)**. Let $\alpha$ be a positive real number and $n$ be an arbitrary integer. Prove that there exists integer pairs (h, k), with $1 < k \leq n$ such that

$$|0 \leq k\alpha - h| < \frac{1}{n}.$$

***Proof*** Let $\{x\} = x - \lfloor x \rfloor$ denote the fractional part of x. It suffices to show that there exists an integer $k \leq n$ such that $0 \leq \{\alpha\} < \frac{1}{n}$. Partition I = [0, 1) into $l_i = \left[\frac{i}{n}, \frac{(i+1)}{n}\right)$ for $0 \leq i \leq n - 1$. Consider the numbers $\{\alpha\}, \{2\alpha\}, \ldots, \{n\alpha\}$. If any of them fall into $l_0$ or $l_{n-1}$, then we are done. Suppose on the contrary that none does, then by the pigeonhole principle, there exists $j < k$, $l_i$ such that both $\{k\alpha\} \in l_i$. But $\{(k - j)\alpha\}$ must fall into $l_0$ or $l_{n-1}$, contradiction. This completes the proof.

***Question*** **(Fermat)**. Show that every prime number of the form $p \equiv 1(\text{mod } 4)$ can be written as a sum of squares of two integers. (i.e. $p = a^2 + b^2$.)

***Proof*** Let p be such a prime, it is simple to show that $\exists x s.t. x^2 \equiv -1(\text{mod p})$. (We will explain these techniques in a later section). It suffices to show that ∃ integer pair (u, v), such that $u + xv \equiv 0(\text{mod p})$ and $|u|, |v| \leq \sqrt{p}$. The desired result would then follow because $u^2 + v^2 \equiv 0(\text{mod p})$, and $0 < u^2 + v^2 < 2p$. Suppose on the contrary that such a pair does not exist. Consider all pairs $(u, v) \neq (0, 0)$, where $-\sqrt{p} \leq u, v \leq \sqrt{p}$. There are at least $(2\lfloor p \rfloor + 1)^2 - 1 \geq 4p$, such pairs. Consider all numbers of the form $u + vx$ (pigeons). Since there $p - 1$ possible residues mod $p$ (holes), by the pigeonhole principle, there must exist pairs $(u, v) \neq (u\prime, v\prime)$, such that $u + xv \equiv u\prime + xv\prime(\text{mod } p)$. But $(|u - u\prime|, |v - v\prime|)$ is a pair satisfying the desired condition. Contradiction.

## 1.8 Recurrence Relations

**efinition** If $n$th expression of a sequence can be considered as a function of the previous expressions

$$x_n = f\left(x_{n-k}, x_{n-k+1}, \ldots, x_{n-1}\right), n > k.$$

This equation is known as the recurrence relation. On the other hand, the initial conditions $x_1, x_2, \ldots, x_k$ must be explicitly determined. Moreover, the function f can also depend on some previous expressions.

In this section we will we will outline some methods of solving recurrence relation. By solving, we mean to find an explicit form of $x_n$ as a function of $n$ that is free of previous terms, except ones given in initial conditions.

In the following, we express some methods to solve the recurrence relation. In other words, we find an explicit form of $x_n$ as a function of $n$ that is independent from previous expressions except for those given in the initial conditions. For instance, the Towers of Hanoi recurrence relation

$$\begin{cases} x_n = 2x_{n-1} + 1 \\ x_1 = 1 \end{cases}$$

has the solution

$$x_n = 2^n - 1.$$

Recurrences are classified according to their structure. Here, we introduce some of them.

- First order:

$$\begin{cases} \text{Linear:} & a_n = 2 \times a_{n-1} + 1 \\ \text{Non - Linear:} & a_n = \frac{1}{1+a_{n-1}} \end{cases}.$$

- Second Order:

$$\begin{cases} \text{Linear:} & a_n = a_{n-1} + a_{n-2} \\ \text{Non - Linear:} & a_n = a_{n-1} \times a_{n-2} \end{cases}.$$

- Higher Order:

$$\begin{cases} \text{Linear:} & a_n = a_{n-1} + a_{n-2} + a_{n-3} \\ \text{Non - Linear:} & a_n = a_0 a_{n-1} + a_1 a_{n-2} + ...a_{n-1} a_0 \end{cases}.$$

- Divide and Conquer

$$\begin{cases} \text{Binary Search:} & a_n = a_{\lfloor \frac{n}{2} \rfloor} + 1 \\ \text{Marge Sort:} & a_n = a_{\lfloor \frac{n}{2} \rfloor} + a_{\lceil \frac{n}{2} \rceil} + n \end{cases}.$$

### *1.8.1 Solving First Order Linear Recurrences*

These iterations can be solved by iterations, also called telescopes. In this method, we repeat until only the initial values remain. In this section, the following classes of linear recurrence relations is considered

$$\begin{aligned} a_n &= \lambda \times a_{n-1} \\ a_n &= \sigma(n) \times a_{n-1} \\ a_n &= \lambda \times a_{n-1} + \tau(n) \\ a_n &= \sigma(n) \times a_{n-1} + \tau(n) \end{aligned}$$

The first two equations and the last two equations are called homogeneous and heterogeneous, respectively.

$a_n = \lambda \times a_{n-1}$

Let us start with the equation

$$a_n = 2 \times a_{n-1}$$
$$a_1 = 1.$$

Then the iteration process is provided as follows:

$$\begin{aligned} a_n &= 2 \times a_{n-1} \\ a_{n-1} &= 2 \times a_{n-2} \\ a_{n-1} &= 2 \times a_{n-3} \\ &\vdots \\ a_2 &= 2 \times a_1. \end{aligned}$$

Applying the back-substitution, we get

$$a_n = 2 \times a_{n-1} = 2^2 \times a_{n-2} = 2^3 \times a_{n-3} = \cdots = 2^{n-1} \times a_1.$$

Therefore,

$$a_n = 2^{n-1}.$$

**heorem** *The recurrence*

$$a_n = \lambda \times a_{n-1}$$

*has the following solution*

$$a_n = \lambda^{n-1} \times a_1.$$

$a_n = \sigma(n) \times a_{n-1}$

**heorem** *The recurrence*

$$a_n = \sigma(n) \times a_{n-1}$$

*has the following solution*

$$a_n = a_1 \prod_{k=2}^{n} \sigma(k).$$

*For instance, if* $\sigma(n) = n$, *we have* $a_n = a_1 \times n!$.

$a_n = a_{n-1} + \tau(n)$

As an example of the inhomogeneous type, we consider

$$a_n = a_{n-1} + n$$
$$a_1 = 1.$$

Using the recurrence to itself.

$$\begin{aligned} a_n &= a_{n-1} + n \\ a_{n-1} &= a_{n-2} + n - 1 \\ a_{n-2} &= a_{n-3} + n - 2 \\ &\vdots \\ a_2 &= a_1 + 2 \end{aligned}$$

Applying the back-substitution, gives

$$a_n = a_{n-1} + n = a_{n-2} + n + (n-1) = a_{n-3} + n + (n-1) + (n-2) = \ldots$$

and we have

$$a_n = a_{n-1} + (n-1) + \cdots + 2 + 1$$
$$a_n = \frac{n(n+1)}{2}.$$

**heorem** *The recurrence*

$$a_n - a_{n-1} + \iota(n), n > 1$$

*has the following solution*

$$a_n = a_1 + \tau(n) + \tau(n-1) + \cdots + \tau(2) + \tau(1)$$
$$a_n = a_1 + \sum_{k=2}^{n} \tau(k).$$

$a_n = \lambda \times a_{n-1} + \tau(n)$

Consider the recurrence relation of Hanoi Towers as follows:

$$a_n = 2 \times a_{n-1} + 1$$
$$a_1 = 1.$$

We proceed in the same way as above. We first use the following iteration

$$\begin{aligned} a_n &= 2 \times a_{n-1} + 1 \\ a_n &= 2 \times a_{n-2} + 1 \\ a_n &= 2 \times a_{n-3} + 1 \\ &\vdots \\ a_n &= 2 \times a_1 + 1 \end{aligned}$$

and then performing back-substitution

$$a_n = 2 \times a_{n-1} + 1 = 2^2 \times a_{n-2} + 2 + 1 = 2^3 \times a_{n-3} + 2^2 + 2 + 1 = \ldots$$

The solution is

$$a_n = 2^{n-1} \times a_1 + 2^{n-2} + 2^{n-1} + \ldots 2 + 1$$

or (since $a_1 = 1$)

$$a_n = 2^{n-1} + 2^{n-2} + 2^{n-1} + \ldots 2 + 1 = 2^n - 1$$

Let us consider the most general case

$$a_n = \lambda a_{n-1} + \tau(n).$$

By iteration, we get

$$\begin{aligned} a_n &= \lambda a_{n-1} + \tau(n) \\ a_n &= \lambda^2 a_{n-2} + \lambda\tau(n-1) + \tau(n) \\ a_n &= \lambda^3 a_{n-3} + \lambda^2\tau(n-2) + \lambda\tau(n-1) + \lambda\tau(n). \end{aligned}$$

The pattern is obvious.

**heorem** *The recurrence*

$$a_n = \lambda a_{n-1} + \tau(n)$$

*has the following solution*

$$a_n = \lambda^{n-1} a_1 + \sum_{k=2}^{n} \lambda^{n-k}(k)\tau.$$

$a_n = \lambda \times a_{n-1} + \tau(n)$

The solution of this state is left to the reader as an exercise.

Note. We have not yet proved the solution of recurrence equations by iterative method. To verify this, we must prove this iterative method by induction.

### 1.8.2 Solving Second Order Recurrences

### 1.8.3 Iteration

In this section, we apply the iteration method to a second order linear recurrence equation. For this purpose, we consider the Fibonacci sequence. In this case, we have

$$\begin{aligned}
a_n &= a_{n-1} + a_{n-2} \\
a_{n-1} &= a_{n-2} + a_{n-3} \\
a_{n-2} &= a_{n-3} + a_{n-4} \\
&-------------- \\
a_n &= a_{n-2} + 2a_{n-3} + a_{n-4}.
\end{aligned}$$

Iterating it twice

$$\begin{aligned}
a_n &= a_{n-2} + 2a_{n-3} + a_{n-4} \\
a_{n-2} &= a_{n-3} + a_{n-4} \\
a_{n-3} &= a_{n-4} + a_{n-5} \\
a_{n-4} &= a_{n-5} + a_{n-6} \\
&----------------- \\
a_n &= a_{n-3} + 3a_{n-4} + 3a_{n-5} + a_{n-6}.
\end{aligned}$$

As can be seen, the iterative method does not work in this case, because the terms on the right do not vanish and thus the pattern does not appear.

#### 1.8.3.1 Characteristic Equation

In this section, we introduce a new method for solving the linear recurrences. For this purpose, we consider the Fibonacci sequence

$$F_n = F_{n-1} + F_{n-2}.$$

From definition, we know that the Fibonacci sequence holds in the following relation

$$F_{n-2} \leq F_{n-1} \leq F_n.$$

Thus, if we replace $F_{n-2}$ term by $F_{n-1}$, we obtain

$$F_n = F_{n-1} + F_{n-2} \leq 2F_{n-1}.$$

On the other hand

$$2F_{n-2} \leq F_{n-1} + F_{n-2} = F_n.$$

Hence,

$$2F_{n-2} \leq F_n \leq 2F_{n-1}.$$

We propose two sequences as follows:

$$b_n = 2 \times b_{n-1}$$
$$c_n = 2 \times c_{n-2}$$

such that

$$c_n \leq F_n \leq b_n.$$

Both equations can be easily solved by iteration

$$b_n = \alpha \times 2^{n-1}$$
$$c_n = \beta \times 2^{\frac{n}{2}}$$

where $\alpha$ and $\beta$ are some constants. Thus, we proved that

$$\beta \times 2^{\frac{n}{2}} \leq F_n \leq \alpha \times 2^{n-1}.$$

Therefore, one can say that the Fibonacci equation has an exponential solution. We set

$$F_n = \lambda^n$$

where $\lambda$ is to be determined. To find $\lambda$, we connect this to the basic equation

$$F_n = F_{n-1} + F_{n-2}$$

In this case, we obtain

$$\lambda^n = \lambda^{n-1} + \lambda^{n-2}.$$

This results a polynomial equation as follows:

$$\lambda^n - \lambda^{n-1} - \lambda^{n-2} = 0$$
$$\lambda^{n-2}\left(\lambda^2 - \lambda - 1\right) = 0$$
$$\lambda^2 - \lambda - 1 = 0.$$

The last equation is known as the characteristic equation. Then, this equation has two roots

$$\lambda_1 = \frac{1-\sqrt{5}}{2}, \lambda_2 = \frac{1+\sqrt{5}}{2}.$$

Thus, solutions of the Fibonacci recurrence is expressed as follows:

$$\left(\frac{1-\sqrt{5}}{2}\right)^n \text{ and } \left(\frac{1+\sqrt{5}}{2}\right)^n.$$

#### 1.8.3.2 General Solution

Let $F_n = \lambda^n$ and $F_n = \lambda^n$ be two solutions. Then their linear combination

$$a_n = c_1 \times \lambda_1^n + c_2 \times \lambda_2^n$$

where $c_1$ and $c_2$ are arbitrary constants is also a solution. Such solution is called a general solution. We prove that a general solution is a solution by substitution $a_n$ into the Fibonacci equation

where the general solution $c_1$ and $c_2$ are arbitrary constants. By plugging $a_n$ in the Fibonacci equation, we prove that a general solution of this equation is obtained by

$$a_n - a_{n-1} - a_{n-2} = 0.$$

We get

$$c_1 \times \lambda_1^n + c_2 \times \lambda_2^n - \left(c_1 \times \lambda_1^{n-1} + c_2 \times \lambda_2^{n-1}\right) - \left(c_1 \times \lambda_1^{n-2} + c_2 \times \lambda_2^{n-2}\right) = 0.$$

Thus,

$$c_1 \times \left(\lambda_1^n - \lambda_1^{n-1} - \lambda_1^{n-2}\right) + c_2 \times \left(\lambda_2^n - \lambda_2^{n-1} - \lambda_2^{n-2}\right) = 0.$$

Each term is zero since $\lambda_1$ and $\lambda_2$ are roots of the characteristic equation. For the Fibonacci sequence, the general solution is

$$a_n = c_1 \times \left(\frac{1-\sqrt{5}}{2}\right)^n + c_2 \times \left(\frac{1+\sqrt{5}}{2}\right)^n.$$

We now obtain $c_1$ and $c_2$ from the initial conditions as follows:

$$a_0 = 0, a_1 = 1.$$

This gives a system as follows:

$$\begin{cases} a_0 = c_1 \times \lambda_1^0 + c_2 \times \lambda_2^0 = 0 \\ a_1 = c_1 \times \lambda_1^1 + c_2 \times \lambda_2^1 = 1 \end{cases}$$

or

$$\begin{cases} c_1 + c_2 = 0 \\ c_1 \times \dfrac{1-\sqrt{5}}{2} + c_2 \times \dfrac{1+\sqrt{5}}{2} = 1 \end{cases}.$$

We obtain

$$c_1 = -\frac{1}{\sqrt{5}}, c_1 = \frac{1}{\sqrt{5}}.$$

Therefore,

$$F_n = -\frac{1}{\sqrt{5}} \times \left(\frac{1-\sqrt{5}}{2}\right)^n + \frac{1}{\sqrt{5}} \times \left(\frac{1+\sqrt{5}}{2}\right)^n.$$

Due to the golden ratio

$$F_n = \frac{\phi^n - (1-\phi)^n}{\sqrt{5}}.$$

It can be easily shown that

$$\lim_{n \to \infty} \frac{F_{n+1}}{F_n} = \phi.$$

***Example*** Solve the following recurrence relation

$$a_n = 3a_{n-1} + 4a_{n-2}$$
$$a_0 = 0,\, a_1 = 1.$$

Suppose the solution is as $a_n = \lambda^n$ and substitute it into the equation to get

$$\lambda^n = 3\lambda^{n-1} + 4\lambda^{n-2}$$
$$\lambda^{n-2}\left(\lambda^2 - 3\lambda + 4\right) = 0.$$

The characteristic equation is expressed as follows:

$$\lambda^2 - 3\lambda - 4 = 0,$$

which has two roots

$$\lambda_1 = -1 and \lambda_2 = 4.$$

The general solution is

$$a_n = c_1 \times (-1)^n + c_2 \times 4^n.$$

We find $c_1$ and $c_2$ from initial conditions. This gives us the following system

$$\begin{cases} a_0 = c_1 + c_2 = 0 \\ a_1 = -c_1 + 4c_2 = 1. \end{cases}$$

Its solution is

$$c_1 = -\frac{1}{5},\, c_2 = \frac{1}{5}.$$

Finally, the solution to the original recurrence is

$$a_n = -\frac{1}{5} \times (-1)^n + \frac{1}{5} \times 4^n.$$

#### 1.8.3.3 Higher Order Equations

For every linear recurrence equation with constant coefficients, the method of the characteristic equation can be applied. It is meant that the coefficients in all $a_k$ expressions of the equation are empty of n. The following equations are constant coefficients

$$a_{n+1} = 2a_{n-1} + 1$$
$$a_n = a_{n-1} + a_{n-2} + a_{n-3}.$$

Consider the following equation

$$a_n = 2a_{n-1} + a_{n-2} - 2a_{n-3}$$
$$a_1 = 0,\ a_2 = 3,\ a_3 = 9.$$

Then, the characteristic equation

$$\lambda^3 - 2\lambda^2 - \lambda + 2 = 0$$

has three roots

$$\lambda_1 = -1,\ \lambda_2 = 1,\ \lambda_3 = 2.$$

The general solution is

$$a_n = c_1(-1)^n + c_2 1^n + c_3 2^n.$$

We find coefficients

$$\begin{cases} a_0 = c_1 + c_2 + c_3 = 0 \\ a_0 = -c_1 + c_2 + 2c_3 = 3 \\ a_1 = c_1 + c_2 + 4c_3 = 9 \end{cases}.$$

Thus, the solution is $a_n = 3(2^n - 1)$.

### 1.8.4 Multiple Roots

Consider the following example

$$a_n = 2a_{n-1} - a_{n-2}$$
$$a_0 = 1,\ a_1 = 2.$$

The characteristic equation

$$\lambda^2 - 2\lambda - 1 = 0$$

has two identical roots

$$\lambda_1 = \lambda_2 = 1.$$

The first solution is $1^n$. But what is the second solution? In general, if all (or several) roots of the characteristic equation are the same, what is the solution to the recurrence? To obtain the concept of the second solution, we consider a new equation

$$b_n = (2 + \varepsilon)b_{n-1} - (1 + \varepsilon)b_{n-2}$$
$$b_0 = 1, b_1 = 2.$$

It's easy to see that if $\varepsilon \to 0$ then the sequence $b_n$ approaches $a_n$. The characteristic equation for $b_n$ sequence is

$$\lambda^2 - (2 + \varepsilon)\lambda - (1 + \varepsilon) = 0.$$

It has two roots

$$\lambda_1 = 1, \lambda_2 = 1 + \varepsilon.$$

Therefore, a general solution is

$$b_n = c_1 \times 1^n + c_2 \times (1 + \varepsilon)^n$$

where $c_1$ and $c_2$ are arbitrary. Consider them in the particular form

$$c_1 = -\frac{1}{\varepsilon}, c_2 = \frac{1}{\varepsilon}.$$

If $\varepsilon \to 0$, we have

$$\lim_{\varepsilon \to 0} \frac{(1 + \varepsilon)^n - 1}{\varepsilon} = \lim_{\varepsilon \to 0} \frac{\left(1^n + n \times \varepsilon \times 1^{n-1} + \cdots + \varepsilon^n\right) - 1}{\varepsilon} = n.$$

This is the second solution for $a_n$ sequence. Then the general solution is

$$a_n = c_1 + c_2 \times n.$$

Taking into account initial conditions

$$a_0 = 1, a_1 = 2$$

we find

$$c_1 = 1, c_2 = 1.$$

Hence, $\lambda_1 = 1, \lambda_2 = 1 + \varepsilon$.

#### 1.8.4.1 More on Multiple Roots

In the previous section we saw that if the characteristic equation has a multiple root $\lambda$ then

$$a_n = \lambda^n \text{ and } a_n = n\lambda^n$$

are solutions. We will prove this directly. Consider the second order recurrence equation

$$a_n = \alpha \times a_{n-1} + \beta \times a_{n-2}.$$

The characteristic equation

$$\lambda^2 - \alpha\lambda - \beta = 0$$

has two identical roots $\lambda_1 = \lambda_2 = \lambda$

$$\lambda_1 = \frac{\alpha - \sqrt{\alpha^2 + 4\beta}}{2}, \lambda_2 = \frac{\alpha + \sqrt{\alpha^2 + 4\beta}}{2}$$

if and only if $\beta = -\frac{\alpha}{2}$. It follows $\lambda = \frac{\alpha}{2}$. To prove that

$$a_n = n\lambda^n$$

is the solution we substitute this into the original recurrence

$$n\lambda^n = \alpha(n-1)\lambda^{n-1} + \beta(n-2)\lambda^{n-2}.$$

Divide this by $\lambda^{n-2}\lambda^{n-2}$

$$n\lambda^2 - \alpha(n-1)\lambda - \beta(n-2) = 0$$

and then collect terms with respect to n

$$n\big(\lambda^2 - \alpha\lambda - \beta\big) + \alpha\lambda + 2\beta = 0.$$

The first term is zero because it is the roots of the characteristic equation. The second term.

is zero because $\beta = -\frac{\alpha^2}{4}$ and $\lambda = \frac{\alpha}{4}$.

**heorem** *Let λ be a root of multiplicity p of the characteristic equation. Then*

$$\lambda^n, n\lambda^n, n^2\lambda^n, ..., n^{p-1}\lambda^n$$

*are all solutions to the recurrence.*

***Example*** Find a general solution

$$a_n = 3a_{n-1} - 3a_{n-2} + a_{n-3}.$$

The characteristic equation has a root $\lambda = 1$ of multiplicity 3. Therefore,

$$a_n = c_1 + c_2 n + c_3 n^2$$

is a solution of this recurrence equation.

### *1.8.5 Inhomogeneous Equations*

As an example of such recurrences, consider the Fibonacci trees. This structure is recursively defined as follows (Fig. 1.17).

- The empty tree is a Fibonacci tree of order 0.
- A single node tree is a Fibonacci tree of order 1.
- A Fibonacci tree of order $n$ has a left Fibonacci subtree of order $n - 1$ and a right Fibonacci subtree of order $n - 2$.

We intend to count the number of nodes of the Fibonacci tree with order n. Let $T_n$ denote the number of tree nodes with order n. In this case

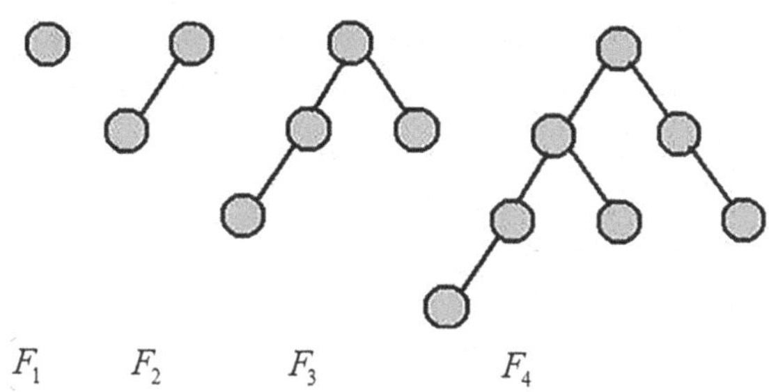


**Fig. 1.17** Geometric representation of Fibonacci tree

$$T_n = T_{n-1} + T_{n-2} + 1$$
$$T_0 = 0,\ T_1 = 1.$$

A recurrence of the form.

$$a_n + \gamma_1 a_{n-1} + \cdots + \gamma_k a_{k-1} = f(n)$$

where all k is constant and $f(n)$ is a non-zero function, is called the linear recurrence equation of inhomogeneous-type with constant coefficients. There is no known general method for solving such equations. We consider a particular case as follows:

$$f(n) = \delta^n p(n)$$

where $p(n)$ is a polynomial and $\delta > 0$. The basic idea is to convert a given heterogeneous equation into a homogeneous equation. For this purpose, we consider the Fibonacci tree recurrence. To cancel on the right hand side, we consider the original equation along with the equation obtained by substituting $n \to n - 1$.

$$T_n - T_{n-1} - T_{n-2} = 1$$
$$T_{n-1} - T_{n-2} - T_{n-3} = 1.$$

Then we subtract the second equation from the first equation

$$T_n - 2T_{n-1} + T_{n-2} = 0$$
$$T_0 = 0,\ T_1 = 1,\ T_2 = 2.$$

This is the homogeneous equation with fourth order that we can solve with the following characteristic equation

$$\left\{\{x \to 1\}, \left\{x \to \frac{1}{2}\left(1 - \sqrt{5}\right)\right\}, \left\{x \to \frac{1}{2}\left(1 + \sqrt{5}\right)\right\}\right\}.$$

The general solution is given by

$$T_n = c_1 + c_2\left(\frac{1 - \sqrt{5}}{2}\right)^n + c_3\left(\frac{1 + \sqrt{5}}{2}\right)^n.$$

The system for coefficients $c_k$

$$T_0 = c_1 + c_2 + c_3 = 0$$
$$T_1 = c_1 + c_2\left(\frac{1-\sqrt{5}}{2}\right) + c_3\left(\frac{1+\sqrt{5}}{2}\right) = 1$$
$$T_2 = c_1 + c_2\left(\frac{1-\sqrt{5}}{2}\right) + c_3\left(\frac{1+\sqrt{5}}{2}\right) = 2$$

has a solution

$$c_1 = -1,\, c_2 = \frac{5-3\sqrt{5}}{10},\, c_2 = \frac{5+3\sqrt{5}}{10}.$$

After some algebra, we get

$$T_n = -1 - \frac{1}{\sqrt{5}}\left(\frac{1-\sqrt{5}}{2}\right)^n + \frac{1}{\sqrt{5}}\left(\frac{1+\sqrt{5}}{2}\right)^n - \frac{1}{\sqrt{5}}\left(\frac{1-\sqrt{5}}{2}\right)^{n+1} + \frac{1}{\sqrt{5}}\left(\frac{1+\sqrt{5}}{2}\right)^{n+1}$$

which can expressed as follows:

$$T_n = F_{n+1} + F_n - 1$$

or

$$T_n = F_{n+2} - 1$$

where $F_n$ is the Fibonacci number.

## 1.9 Exercises

1. Consider the recurrence

$$a_1 = 2 \text{and} a_n = a_{n-1}^2 + 1, \text{for} n \geq 2.$$

   Write down the first five terms of this sequence.

2. Give a formula for the general term of

   (a) The sequence 2, 4, 6, 8, ... of even numbers,
   (b) The sequence 1, 4, 9, 16, ... of squares.

3. Find the $n$th term of the arithmetic sequence $\log_5^2, \log_5^4, \log_5^8, \ldots$
4. Show that 12 is not a term of the arithmetic sequence 210, 197, 184, ...
5. Find the $n^{th}$ term of the geometric sequence $\sqrt{6}, 2\sqrt{3}, 2\sqrt{6}, \ldots$
6. By writing out the terms, find the sum

$$\sum_{k=1}^{n}\left(\frac{1}{k}-\frac{1}{k+1}\right).$$

7. Use the method of writing the arithmetic series

$$a+(a+d)+(a+2d)+\ldots(\ell-d)+\ell$$

forwards and backwards to derive the formula $S_n = \frac{n}{2}(a+\ell)$ given above.

8. Sum the arithmetic series

$$\log_2^3+\log_2^9+\log_2^{27}+\ldots$$

to $n$ terms.

9. Use the method of multiplying the geometric series

$$a+ar+ar^2+\cdots+ar^{n-1}$$

by $r$ and subtracting to derive the formula for $S_n$ given above.

10. Find the sum to $n$ terms of the geometric series $\sqrt{3}+6+12\sqrt{3}+\ldots$
11. Suppose $a$ and $b$ are positive real numbers. First, take a line segment $AB$ of length $Ab$ and mark a point $X$ on it such that $AX = a$ (and so $XB = b$). Next, draw a semicircle, centered at the midpoint of $AB$, with $AB$ as diameter. Finally, raise a perpendicular to $AB$ at $X$ to meet the semicircle at $Y$.
    (a) Prove that the length $XY$ is the geometric mean of $a$ and $b$.
    (b) By noting that the radius of the semicircle equals the arithmetic mean of $a$ and $b$, deduce that

$$\frac{a+b}{2} \geq \sqrt{ab}.$$

That is, the arithmetic mean of $a$ and $b$ is greater than or equal to their geometric mean.

12. Explain why the geometric series

$$1+\frac{1}{1+\sqrt{2}}+\left(3-2\sqrt{2}\right)+\ldots$$

has a limiting sum, and find its value.

13. By writing the recurring decimal $0.\overline{12} = 0.121212...$ as

$$\frac{12}{10^2} + \frac{12}{10^4} + ...,$$

express $0.\overline{12}$ as a rational number in simplest form.

14. Suppose that $a_1, a_2, ..., a_n$ is a sequence of positive real numbers and b > 1. Show that the logarithm (base b) of the geometric mean of these numbers is equal to the arithmetic mean of the numbers $\log_b^{a_1}, \log_b^{a_2}, ..., \log_b^{a_n}$.

15.

- Find the arithmetic, geometric and harmonic means of 3, 4, 5 and write them in ascending order.
- Prove that the harmonic mean of two positive real numbers a and b is less than or equal to their geometric mean.

16.

(a) Find the maximum value of the function $f(x) = \log_e^x - x$, for $x > 0$. Hence, deduce that $f(x) = \log_e^x \leq x - 1$, for all $x > 0$.

(b) Let $a_1, a_2, ...a_n$ be positive real numbers and define $A = \frac{a_1.a_2....a_n}{n}$. By successively substituting $x = \frac{a_i}{A}$, for $i = 1, 2, ..., n$, into the inequality from part (a) and summing, show that

$$\log_e^{\left(\frac{a_1 a_2 ... a_n}{A^n}\right)} \leq 0.$$

(c) By exponentiating both sides of the inequality from part (b), derive the generalized AM–GM inequality.

17. Show that every integer $N \neq 0$ in the form where $k$ is a non-negative integer and $l$ is an odd integer.

18. Let $\{a_i\}$ be a sequence of real numbers that satisfy $a_{i+j} \leq a_i + a_j, \forall i, j$. Prove that

$$\frac{a_1}{1} + \frac{a_2}{2} + \cdots + \frac{a_n}{n} \geq a_m.$$

19. Prove that every positive integer n has a binary expression. Namely, that there exists integers $c_i \in \{0, 1\}$ such that

$$n = c_r 2^r + c_{r-1} 2^{r-1} + \cdots + c_2 2^2 + c_1 2^1 + c_0 2^0.$$

20. Consider the sequence defined as $d_1 = 1, d_2 = 2, d_3 = 3$, and $d_{n+3} = d_{n+1} + d_n$ for all positive integers $n$. Show that $d_n < 2^n$.

21. The real numbers $x_1, x_2, x_3, ...$ satisfy $x_{i+j} \leq x_i + x_j$ for all $i, j$. Prove that

$$x_1 + \frac{x_2}{2} + \cdots + \frac{x_n}{2} \geq x_n.$$

22. For $\geq 1$, prove that

- $\sum_{i=1}^{n} i^2 = \frac{n(n+1)(2n+1)}{6}$
- $\sum_{i=1}^{n} k!k = (n+1)! - 1$

23.

- Use direct poof to prove that if $a - b$ is even, then $a^2 - b^2$ is even.
- Prove by contradiction that $\log_5(3)$ is irrational value.
- Use indirect proof to prove that if $n^2 > 6n$ then $n \neq 4$.
- Prove or disprove the following statement: for every natural number $n$, the integer $4n^3 - 6n^2 + 141$ is a prime number.
- Prove that for some irrational numbers $x$ and $y$, $x^y$ is rational.
- Prove that if $2 = -3$ then man can run faster than light speed.

24. Prove by induction that for any integer $n > = 2$ the following numbers

$$u_2 = \sqrt{1 + \sqrt{1}}, u_3 = \sqrt{1 + \sqrt{1 + \sqrt{1}}}, \ldots, , u_\infty = \sqrt{1 + \sqrt{1 + \sqrt{1 \ldots}}}$$

are irrational.

25.

- Use direct poof to prove that if $n$ is an even integer then $n^2$ is even integer also.
- Prove by contradiction that $\log_4(5)$ is irrational value.
- Use indirect poof, contrapositive, to prove that if $n + m > 100 \to m > 50$ or $n > 50$
- Prove if $x^2 + 1 < 0$, then $x^3 > 4$.

26.

- For $n > 1$, prove that

$$\left(1 - \frac{1}{2}\right)\left(1 - \frac{1}{3}\right) \ldots \left(1 - \frac{1}{n}\right) = \frac{1}{n}.$$

- Let $a_n$ be a sequence defined as follows:

$$\left\{ a_{n+1} = \frac{1}{2}\left(a_n + \frac{2}{a_n}\right) a_1 = 1 \right.$$

(a) For all $n > 1$ prove that $a_n \geq \sqrt{2}$.
(b) Conclude that $a_{n+1} - a_n \leq 0$.
(c) Find the limit of sequence, $a_n$.

- Let $a_n$ be a sequence defined as follows:

$$\{a_{n+1} = 5a_n - 6a_{n-1}a_1 = a_2 = 1$$

(a) Calculate $a_{100}$.
(b) Calculate $a_n$.

$$2x + 6y = 1.$$

29. Prove by contrapositive that for $n > 0$

    ($4^n - 1$ is prime) ($n$ is odd).
    Hint: $A^2 - B^2 = (A - B)(A + B)$.

30. Disprove that: If $12|x^2$, then $12|x$.
31. Prove that $\sqrt{p}$ is irrational for every prime $p$.
32. Prove that there is a prime number between 90 and 100.
33. Prove that there are two irrational numbers $\alpha, \beta$ such that $\alpha^\beta$ is rational.
34. Find all $k$ such that $(k - 2)|2k$.

    Hint: what else does $(k - 2)$ divide?

35. Let $d = gcd(a, b)$.
    - Prove, without using prime factorization, that if $a|n$ and $b|n$, then $ab|dn$.
    - Prove, without using prime factorization, that if $c|a$ and $c|b$, then $c|d$.
36. The least common (LCM) of natural numbers $a, b$ is the least natural number that is divisible by both. Prove that $lcm(a, b). \text{gcd}(a, b) = a \times b$.
37. Let $a = 300$ and $b = 135$
    - Use the Euclidean algorithm to find gcd(300, 135).
    - Write the GCD as an integer combination of 300 and 135.
    - Find all integer solutions to the Diophantine equation $300x + 135y = -45$.
38. Find all solutions of in $x^{n+1} - (x + 1)^n = 2001$ positive integers $x$ and $n$.
39. Solve the following congruences
    - $5x \equiv 8^{26} (\text{mod} 13)$
    - $x \equiv 5^{183} (\text{mod} 99)$.
40. For which natural numbers $n$ is the number $n^2 - 2016$ a perfect square?
41. Let $q$ be a prime factor of $a^2 + b^2$. Show that if $q \equiv 3(\text{mod} 4)$, then $q|a$ and $q|b$.
42. The sequence $a_n$ is defined by $a_1 = 2$, $a_{n+1} = a_n^2 - a_n + 1$. Show that any pair of values in the sequence are relatively prime.

43. Argue that $p \not| (r_1 \times r_1)$ if $r_1 \times r_2 \neq 0$. Note $p$ is prime.
44. Where will the proof fail if you try to prove that $\sqrt{4}$ is irrational?
45. Draw a Hasse diagram for $p^k$ where p is prime.
46. How many positive integers less than 100 have exactly 6 divisors?
47. Prove that

$$\gcd(a, b) = \gcd(b, a - b).$$

Given a pair of numbers, does a gcd always exist? Is it unique?

48. When does the worst performance of the Euclidean Algorithm occur?
49. Apply the Euclidean algorithm to two real numbers. What will this procedure yield?
50. How do you find such n, m that $\gcd(a, b) = n \times a + m \times b$?
51. Find a particular solution of

$$56x + 72y = 40.$$

52. Find all integer solutions of

$$56x + 72y = 40.$$

53. Determine the number of solutions in positive integers

$$4x + 7y = 117.$$

54. Find all solution of the following equation

$$3x + 5y = 2.$$

55. Let us choose denominations 3 and 5. What is the largest number that cannot be decomposed?
56. Choose denominations 4 and 5. What is the largest number that cannot be decomposed?
57. How about this set of denominations: 5 and 7. What is the largest number that cannot be decompose?
58. Prove combinatorially (without using the above theorem) that

$$C(n, k) = C(n - 1, k) + C(n - 1, k - 1).$$

59. Solve the recurrence

$$a_n = \frac{n}{n+1} a_{n-1}$$
$$a_1 = 1.$$

60. Solve the recurrence

$$a_n = 2a_{n-1} + 2^n$$
$$a_0 = 1.$$

62. 61 Solve the recurrence

$$a_n = \frac{n}{n+1} a_{n-1} + n^2$$
$$a_0 = 1.$$

62. Solve the recurrence

$$a_n - 5a_{n-1} + 7a_{n-2} - 3a_{n-3} = 0$$
$$a_0 = 1,\ a_1 = 2,\ a_2 = 3.$$

## Reference

1. Grigas A (2013) The Fibonacci sequence: its history, significance, and manifestations in nature, digitalcommons.liberty.edu.

# Chapter 2
# Probability Theory and Applications

Probability is the mathematical language for quantifying uncertainty. We can apply probability theory to a diverse set of problems, from coin flipping to the analysis of computer algorithms. The starting point is to specify a sample space, the set of possible outcomes.

## 2.1 Sample Space and Events

The sample space $\Omega$, is the set of possible outcomes of an experiment. Points $w$ in $\Omega$ are called sample or realizations. Events are subsets of $\Omega$.

***Example*** If we toss a coin twice then $\Omega = \{HH, HT, TH, TT\}$. The event that the first toss is heads is $A = \{HH, HT\}$.

***Example*** Let $w$ be the outcome of a measurement of some physical quantity, for example, temperature. Then $\Omega = \mathbb{R} = (-\infty, \infty)$. The event that the measurement is larger than 10 but less than or equal to 28 is $A = (10, 23]$.

Let $E$ be the event that the first head appears on the third toss. Then

$$E = \{(w_1, w_2, w_3, ...) :\ w_1 = T.w_2 = T.w_3 = H, w_i \in \{H, T\}\, for\ i > 3\}.$$

Given an event $A$, let $A^c = \{w \in \Omega;\ \ w \notin A\}$ denote the complement of $A$. Informally, $A^c$ can be read as "not $A$." The complement of $\Omega$ is the empty set 0. The union of events $A$ and $B$ is defined

$$A \cup B = \{w \in \Omega; w \in A\, or\, w \in B\, or\, w\ \in\ both\}$$

which can be thought of as "$A$ or $B$". If $A_1, A_2, ...$ is a sequence of sets, then



S. Brahim Belhaouari et al., *Discrete Mathematics, Probability Theory and Stochastic Processes*, Modeling and Optimization in Science and Technologies 20,
https://doi.org/10.1007/978-3-031-80589-9_2

**Table 2.1** The notation of sample space, events and operators on them

| | |
|---|---|
| $\Omega$ | Sample space |
| $\omega$ | Outcome |
| $A$ | Event |
| $\|A\|$ | Number of points in $A$ (if $A$ is finite) |
| $A^c$ | Complement of $A$ (not $A$) |
| $A \cup B$ | Union ($A$ or $B$) |
| $A \cap B$ | Intersection ($A$ and $B$) |
| $A - B$ | Set difference (points in $A$ that are not in $B$) |
| $A \subset B$ | Set inclusion ($A$ is a subset of or Equal to $B$) |
| $\emptyset$ | Null event (always false) |

$$\bigcup_{i=1}^{n} A_i = \{w \in \Omega;\ w \in A_i \textit{ for at least one } i\}.$$

The intersection of $A$ and $B$ is $A \cap B = \{w \in \Omega;\ w \in A \textit{ and } w \in B\}$ read "$A$ and $B$." Sometimes we write $A \cap B$ as $AB$. If $A_1, A_2, \ldots$ is a sequence of sets then

$$\bigcap_{i=1}^{n} A_i = \{w \in \Omega\,;\, w \in A_i \textit{ for all } i\}.$$

Let $A - B = \{w \in \Omega;\ w \in A, w \notin B\}$. If every element of $A$ is also contained in $B$, then we write or, equivalently, $B \supset A$. If $A$ is a finite set, let $|A|$ denote the number of number of element in $A$. See Table 2.1 for a summary.

We say that $A_1, A_2, \ldots$ are disjoint or are mutually exclusive if $A_i \cap A_j = \emptyset$ whenever $i \neq j$. For example, $A_1 = [0, 1), A_1 = [1, 2), A_1 = [2, 3), \ldots$ are disjoint. A partition of is a sequence of disjoint sets $A_1, A_2, \ldots$ such that $\bigcup_{i=1}^{\infty} A_i = \Omega$.. Given an event $A$, define the indicator function of $A$ by

$$I_A(w) = I(w \in A) = \begin{cases} 1 \textit{ if } w \in A \\ 0 \textit{ if } w \notin A \end{cases}.$$

A sequence of sets $A_1, A_2, \ldots$ monotone increasing if $A_1 \subset A_2 \subset \ldots$ and we define $\lim_{n\to\infty} A_n = \bigcup_{i=1}^{\infty} A_i$. A sequence of sets $A_1, A_2, \ldots$ is monotone decreasing if $A_1 \supset A_2 \supset \ldots$ and then we define $\lim_{n\to\infty} A_n = \bigcap_{i=1}^{\infty} A_i$. In either case, we will write $A_n \to A$.

***Example*** Let $\Omega = \mathbb{R}$ and let $A_i = [0, \frac{1}{i})$ for $i = 1, 2, ....$ Then $\bigcup_{i=1}^{\infty} A_i = [0, 1)$ and $\bigcap_{i=1}^{\infty} A_i = \{0\}$. If instead we define $A_i = \left(0, \frac{1}{i}\right)$ then $\bigcup_{i=1}^{\infty} A_i = (0, 1)$ and $\bigcap_{i=1}^{\infty} A_i = \emptyset$.

## 2.2 Probability

Probability is a mathematical measure of the likelihood of an event occurring. It is expressed as a number between 0 and 1, where 0 indicates that the event will not happen, and 1 indicates that the event will definitely happen. The higher the probability, the more likely the event is to occur. One common way to define the probability of an event A is through the concept of a probability measure. In probability theory, a probability measure is a function that assigns each event $A$ a real number $P(A)$. For this purpose, we want to assign a real number $P(A)$ to every event $A$, call the probability of $A$. We also call $P$ a probability distribution or a probability measure. To qualify as a probability, $P$ has to satisfy three axioms:

**Definition** A function $P$ that assigns a real number $P(A)$ to each event $A$ is a probability distribution or a probability measure if it satisfies the following three axioms:

- Axioms 1. $\mathrm{P}(A) \geq 0$ for every $A$.
- Axiom 2. $\mathrm{P}(\Omega) = 1$.
- Axiom 3. If $A_1, A_2, ...$ are disjoint then,

$$\mathrm{P}\left(\bigcup_{i=1}^{\infty} A_i\right) = \sum_{i=1}^{\infty} P(A_i).$$

Frequencies and degrees of belief are two common interpretations of $P(A)$. In frequency interpretation, $P(A)$ is the number of times that $A$ is true in repetitions. For instance, if the probability of heads is 1/2, it means that if we flip the coin a few times, as the number of those flips increases, the ratio of the number of times we return will be 1/2. Unpredictable sequence that is infinitely long and has a limiting ratio to a constant is idealized. The interpretation of the degree of belief is that $P(A)$ measures the power of the belief that observes $A$ correctly. In both interpretations, we need principles 1 to 3 to be established. As long as we do not make a statistical inference once, the difference in interpretation will not matter much. Different interpretations lead to two schools of inference: the repeated school and the Bayesian concept. We postpone the discussion.

One can derive many properties of $P$ from the axioms. Here are a few:

$$\begin{aligned} P(\emptyset) &= 0 \\ A \subset B &\Rightarrow P(A) \leq P(B) \\ 0 \leq P(A) &\leq 1 \\ P(A^c) &= 1 - P(A) \\ A \cap B &= \emptyset \Rightarrow P(A \cup B) = P(A) + P(B). \end{aligned}$$

A less obvious property is given in the following Lemma.

**Lemma** *For any events A and B,* $P(A \cup B) = P(A) + P(B) - P(AB)$.

***Proof*** Write $A \cup B = (AB^c) \cup (A^c B)$ and note that these events are disjoint. Hence, making repeated use of the fact that $P$ is additive for disjoint events, we see that

$$\begin{aligned} P(A \cup B) &= P((AB^c) \cup (AB) \cup (A^c B) \\ &= P(AB^c) + P(AB) + P(A^c B) \\ &= P(AB^c) + P(AB) + P(A^c B) + P(AB) - P(AB) \\ &= P\big((AB^c) \cup (AB)\big) + P\big((A^c B) \cup (AB)\big) - P(AB) \\ &= PA) + P(B) - P(AB). \end{aligned}$$

***Example*** **(Two coin tosses)**

Let $H_1$ be the event that heads occurs on toss 1 and let $H_2$ be the event that heads occurs on toss 2. If all outcomes are equally likely, that is

$$P(\{H_1, H_2\}) = P(\{H_1, T_2\}) = P(\{T_1, H_2\}) = P(\{T_1, T_2\}).$$

Then

$$P(H_1 \cup H_2) = P(H_1) + P(H_2) - P(H_1 H_2) = \frac{1}{2} + \frac{1}{2} - \frac{1}{4} = \frac{3}{4}.$$

**Theorem** (Continuity of probabilities).

*If* $A_n \to A$ *then* $P(A_n) \to P(A)$ *as* $n \to \infty$.

***Proof*** Suppose that $A_n$ is monotone increasing so that $A_1 \subset A_2 \subset \ldots$ Let $A = \lim\limits_{n\to\infty} A_n = \bigcup\limits_{i=1}^{\infty} A_i$. Define.

$$B_1 = A_1, B_2 = \{w \in \Omega : w \in A_2, w \notin A_1\}, B_3 = \{w \in \Omega : w \in A_3, w \notin A_2, w \notin A_1\}, \ldots$$

It can be shown that $B_1, B_2, \ldots$ are disjoint, $A_n = \bigcup\limits_{i=1}^{n} A_i = \bigcup\limits_{i=1}^{n} B_i$ for each $n$ and $\bigcup\limits_{i=1}^{n} B_i = \bigcup\limits_{i=1}^{n} A_i$. From Axiom 3,

$$P(A_n) = P\left(\bigcup_{i=1}^{n} B_i\right) = \sum_{i=1}^{n} P(B_i)$$

and, hence, using Axiom 3 again,

$$\lim_{n\to\infty} P(A_n) = \lim_{n\to\infty} \sum_{i=1}^{n} P(B_i) = \sum_{i=1}^{\infty} P(B_i) = P\left(\bigcup_{i=1}^{\infty} B_i\right) = P(A).$$

### 2.2.1 Probability on Finite Sample Spaces

Assume that $\Omega = \{w_1, ..., w_n\}$ is a finite sample space. For instance, if we toss a die twice, then $\Omega$ has 36 elements: $\Omega = \{(i, j) : i, h \in \{1, ..., 6\}\}$. If each outcome is equally likely, then $P(A) = |A|/36$ where $|A|$ denotes the number of elements in $A$. The probability that the sum of the dice is 11 is 2/36 since there are two outcomes that correspond to this event.

In general, if $B_1, B_2, ...$ is finite and if each outcome is equally likely, then

$$P(A) = \frac{|A|}{|\Omega|},$$

which is called the uniform probability distribution. To compute probabilities, we need to count the number of points in an event $A$. Methods for counting points are called combinatorial methods. We needn't delve into these in any great detail. We will, however, need a few facts from counting theory that will be useful later. Giving $n$ object, the number of ways of ordering these objects is $n! = n(n-1)(n-2)\ldots 3.2.1.$ for continence, we define $0! = 1$.

### 2.2.2 Independent Events

If we flip a fair coin twice, then the probability of two heads is $\frac{1}{2} \times \frac{1}{2}$. We multiply the probabilities because we regard the two tosses as independent. The formal definition of independent is as follows:

**Definition** Two events $A$ and $B$ are independent if.

$$P(AB) = P(A)P(B),$$

and we write $AIIB$. A set of events $\{A_i : i \in I\}$ is independent if

$$P\left(\bigcap_{i\in J} A_i\right) = \prod_{i\in J} P(A_i)$$

for every finite subset $J$ of $I$.

Independence can be created in two different tracks. We suppose that two events are independent. For instance, when tossing a coin twice, we suppose that the throws are independent, which indicates the coin has no retention of the first toss. In another example, by affirming it, we gain independence. Assume that $P(AB) = P(A)P(B)$ holds. For example, in tossing a fair die, let $A = \{2, 4, 6\}$ and let $B = \{1, 2, 3, 4\}$. Then,

$$A \cap B = \{2, 4\}, P(AB) = \frac{2}{6} = P(A)P(B) = \frac{1}{2} \times \frac{2}{3}$$

and so $A$ and $B$ are independent. In this case, we didn't assume that $A$ and $B$ are independent it just turned out that they were.

Suppose that $A$ and $B$ are disjoint events, with positive probability. Can they be independent? No. This follows since $P(A)P(B) \geq 0$ yet $P(AB) = P(\emptyset) = 0$. Except in this special case, there is no way to judge independence by looking the sets in a Venn diagram.

***Example*** Toss a fair coin 10 times. Let $A =$"at least one Head". Let $T_j$ be the event that tail occurs on the j-th toss. Then

$$\begin{aligned}
P(A) &= 1 - P(A^c) \\
&= 1 - P(alltails) \\
&= 1 - P(T_1 T_2 ... T_{10}) \\
&= 1 - P(T_1)P(T_2)...P(T_{10}) \\
&= 1 - \left(\frac{1}{2}\right)^{10} \approx 0.999.
\end{aligned}$$

***Example*** Two people take turns trying to sink a basketball into a net. Person 1 succeeds with probability 1/3 while person 2 succeeds with probability 1/4. What is the probability that person 1 succeeds before person 2? Let E denosste the event of interest. Let $A_j$ be the event that the first Success is by person 1 and that it occurs on trial number $A =$ Note that $A_1, A_2, ...$ are disjoint and that $E = \bigcup_{j=1}^{\infty} A_j$. Hence,

$$P(E) = \sum_{j=1}^{\infty} P(A_j).$$

Now, $P(A_1) = 1/3$. $A_2$ occurs if we have the sequence 1 misses, 2 misses, 1 succeeds. This has probability $P(A_2) = (2/3)(3/4)(1/3) = (1/2)(1/3)$. Following this logic we see that $P(A_j) = (1/2)^{j-1}(1/3)$. Hence

$$P(E) = \sum_{j=1}^{\infty} \frac{1}{3}\left(\frac{1}{2}\right)^{j-1} = \frac{1}{3}\sum_{j=1}^{\infty}\left(\frac{1}{2}\right)^{j-1} = \frac{2}{3}.$$

Here we use that fact that, if $0 < r < 1$, then $\sum_{j=k}^{\infty} r^j = r^k/(1-r)$.

**Conditional probability**

Assuming that $P(B) > 0$, we define the conditional probability of $A$ given that $B$ has occurred as follows:

**Definition** If $P(B) > 0$, then the conditional probability of $A$ given $B$ is

$$P(A|B) = \frac{P(AB)}{P(B)}.$$

Think of $P(A|B)$ as the fractional of times $A$ occurs among those in which $B$ occurs. Here are some facts about conditional probabilities. For any fixed $B$ such that $P(B) > 0$, $P(.|B)$ is a probability i.e. it satisfies the third axioms of probability. In particular, $P(A|B) \geq 0$, $P(\Omega|B) = 1$ and if $A_1, A_2, ...$ are disjoint then

$$P\left(\bigcup_{i=1}^{\infty} A_i|B\right) = \sum_{i=1}^{\infty} P(A_i|B).$$

It is not true generally to say that $P(A|B \cup C) = P(A|B) + P(A|C)$. is always valid. For instance, the likelihood of contracting the measles is one, but it is not one if you already have the disease.

In this case, the difference between $P(A|B)$ and $P(B|A)$ is obvious, but there are reasons that are less explicit. This error is made so much in civil reasoning that it is sometimes known as the prosecutor's wrong.

***Example*** A medical test for a disease $D$ has outcomes $+$ and $-$. The probabilities are:

| | $D$ | $D^c$ |
|---|---|---|
| + | 0.0081 | 0.0090 |
| − | 0.0900 | 0.9010 |

From the definition of conditional probability,

$$P(+|D) = \frac{P(+, D)}{P(D)} = \frac{0.0081}{0.0081 + 0.0009} = 0.9$$

and

$$P(-|D) = \frac{P(-, D^c)}{P(D^c)} = \frac{0.9010}{0.9010+0.0900} \approx 0.9.$$

The test appears to be relatively accurate. Bedfast people have a positive return 90% of the time and healthy people have a negative performance 90% of the time. Suppose you leave for a test and your test is positive. What is the probability of getting this disease? Most people answer 90%. The true answer is

$$P(D|+) = \frac{P(+, D)}{P(+)} = \frac{0.0081}{0.0081 + 0.0900} = 0.08.$$

The lesson here is that you need to compute the answer numerically. Don't trust your intuition.

If $A$ and $B$ are independent events, then

$$P(A|B) = \frac{P(AB)}{P(B)} = \frac{P(A)P(B)}{P(B)} = P(A).$$

So another interpretation of independent is that knowing $B$ doesn't change the probability of $A$. From the definition of conditional probability we can also write

$$P(AB) = P(A|B)P(B)$$

and also

$$P(AB) = P(B|A)P(A).$$

Often, these formulae given us a convenient way to compute $P(AB)$ when $A$ and $B$ are not independent.

***Example*** Draw two cards from a deck, without repayment. Let $A$ be the events that the first draw is the ace of clubs and let $B$ be the events that the second draw is the queen of diamonds. Then

$$P(A, B) = P(A)P(B|A) = \frac{1}{52} \times \frac{1}{51}.$$

### 2.2.3 Bayes' Theorem

Bayes' Theorem is a way to update the probability of a hypothesis or event based on new evidence or information. It allows us to take into account prior knowledge or beliefs about the likelihood of a certain event, and adjust that belief as new information becomes available. Essentially, it helps us make more informed decisions by combining our initial beliefs with new data to arrive at a more accurate probability estimate. Bayes' theorem is a useful result that is the basis of "expert system" and "Bayes' nets". First, we need a preliminary result.

**Theorem** (The law of total probability)

*Let $A_1, \ldots, A_k$ be a partition of $\Omega$. Then, for any event B,*

$$P(B) = \sum_{i=1}^{k} P(B|A_i)P(A_i).$$

***Proof*** Define $C_j = BA_j$ and note that $C_1, \ldots, C_k$ are disjoint and that $B = \bigcup_{j=1}^{k} C_j$. Hence,

$$P(B) = \sum_j P(C_j) = \sum_j P(BA_j) = \sum_j P(B|A_j)P(A_j)$$

since $P(BA_j) = P(B|A_j)P(A_j)$ from the definition of conditional probability.

**Theorem** (Bayes' Theorem)

*Let $A_1, \ldots, A_k$ be a partition of $\Omega$ such that $P(A_i) > 0$ for each i. if $P(B) > 0$ then, for each $i = 1, \ldots, k$,*

$$P(A_i|B) = \frac{P(B|A_i)P(A_i)}{\sum_j P(B|A_j)P(A_j)}.$$

***Remark*** We call $P(A_i)$ the prior probability of $A$ and $P(A_i|B)$ the posterior probability of $A$.

***Proof*** We apply the definition of conditional probability twice. Followed by the law of total probability:

$$P(A_i|B) = \frac{P(A_iB)}{P(B)} = \frac{P(B|A_i)P(A_i)}{P(B)} = \frac{P(B|A_i)P(A_i)}{\sum_j P(A|A_j)P(A_j)}.$$

***Example*** I divide my email into three categories: $A_1 =$ "*spam*", $A_2 =$ "*low priority*" and $A_3 =$ "*high priority*". From previous experience I find that $P(A_1) = 0.7, P(A_2) = 0.2$ and $P(A_3) = 0.1$. Of course, $0.7+0.2+0.1 = 1$. Let $B$ be the events that the email contains the word "free". From previous experience, $P(B|A_1) = 0.9, P(B|A_2) = 0.01$ and $P(B|A_3) = 0.01$. (Note: $0.9+0.01+0.01 \neq 1$) I receive an email with the word "free". What is the probability that it is spam?

Solution

Bayes' Theorem yields,

$$P(A_1|B) = \frac{0.9 \times 0.7}{(0.9 \times 0.7) + (0.01 \times 0.2) + (0.01 \times 0.1)} = 0.995.$$

## 2.3 Random Variables

A random variable is a mathematical quantity whose value is determined by the outcome of a random event. In other words, it is a variable whose value is subject to chance or unpredictability. Random variables can be classified as either discrete or continuous. Discrete random variables take on a finite number of values, such as the number of heads in a series of coin flips. Continuous random variables, on the other hand, can take on any value within a certain range, such as the height of a person. Random variables can be used to model various real-world phenomena, such as the outcome of a dice roll, the success or failure of a marketing campaign, or the time it takes for a customer to complete a purchase. In probability theory and statistics, random variables play a key role in analyzing and interpreting data, making predictions, and making decisions in situations where there is uncertainty or randomness.

Since events are considered subsets of event space $\Omega$, to calculate their probability we have to deal with calculations on sets, which of course is not an easy task. Instead, the definition of "random variable" can be used to calculate the probability of many events based on available probabilistic patterns, because many random phenomena have specific patterns.

With the help of a random variable, a number from real space is considered for each event of the sample space. Since events occur randomly, it is natural to consider a random value for each. For this reason, such variables are called random variables.

Suppose in a random experiment with sample space $\Omega$, we denote every simple event by $\omega$. Thus if we consider $X$ to be a real number for the event, we write $X(\omega) \in \mathbb{R}$, which means $\mathbb{R}$ is a set of real numbers.

Now, we call $X$ a random variable if, for any interval from real numbers like $B$, we can show that $X(\omega) \in B$ can be converted to an event, that is, it belongs to the event space. More precisely, if we consider any set like $B$ as a subset of real numbers, then $X^{-1}(B) = \{\omega | X(\omega) \in B\}$ must be an event and belong to $\Omega$. In other words, if we consider $X^{-1}$ to be the inverse of the operation that $X$ performed, for every Borel set of real numbers, we can find an event in the event space that is converted by $X$ to a subset of real numbers (such as $B$). A "Borel Set" is a subset of real numbers that can be represented as open interval $(a, b)$.

As we can see, the random variable $X$ is actually a function of the event space to a set of subsets of real numbers. This means that the random variable $X$ converts the members of the sample space to real numbers, the members of the event space to the Borel set $B$ of real numbers, and the event-related probability function to the probability of the random variable.

**Definition** In probability and statistics, a random variable, random quantity, aleatory variable, or stochastic variable is described informally as a variable whose values depend on outcomes of a random phenomenon. The formal mathematical treatment of random variables is a topic in probability theory. In that context, a random variable is understood as a measurable function defined on a probability space that maps from the sample space to the real numbers.

Symbolically, assume that $X$ is a random variable and $\Omega$ the sample space. For thereal number $x$, if $w \in \Omega$, then $X(w) = x$. The domain $X$ is the sample space and its range is the set of possible values.

The values of the random variable can be discrete, like integers (limited or infinite), continuous, like distances on a real line, or mixed (discrete and continuous). For instance, the values of random variables may be intervals and natural numbers as depictions of certain events.

Notice that the random variable range is specified before performing a randomized experiment, while the random variable value is specified after the test.

**Definition** A discrete random variable means a function (or a map), such as $X$, from a sample space $\Omega$, to a set of real numbers. Symbolically, for real number $x$, if $w \in \Omega$ then $X(w) = x$.

If a random variable can take only a finite number of discrete values, then it is discrete.

***Example*** In tossing a coin, the sample space is $\Omega = \{H, T\}$, where $H$ and $T$, the outcomes of the random experiment of tossing, stand for "heads" and "tails", respectively. In this case, the discrete random variable $X$ can take values 0 and 1 (or other numbers) for $T$ and $H$. Thus we can write:

$$X = \begin{cases} H \to 1 \\ T \to 0 \end{cases}.$$

Similar to above example, we toss two coins. In this case, a set of feasible results, that is, the sample space $\Omega = \{(T, T), (H, H), (T, H), (H, T)\}$. Therefore, the elements of the sample space this time is a set of ordered pairs. Thus, we define a discrete random variable $X$ such that it assigns a number to each of the ordered pair in $\Omega$ as follows:

$$X = \begin{cases} (H, H) \to 1 \\ (H, T) \to 2 \\ (T, H) \to 3 \\ (T, T) \to 0 \end{cases}.$$

***Example*** Consider a cube object with numbers 1 to 6 written on its each side without duplication. Therefore, if this cubic tosses, the set of these numbers is the sample of this experiment.

Consider throwing a dice pair. In this case, the 36 possible results of this experiment will be as follows:

$$\Omega = \begin{Bmatrix} (1, 1), (1, 2), (1, 3), (1, 4), (1, 5), (1, 6), (2, 1), (2, 2), (2, 3), (2, 4), (2, 5), (2, 6), \\ (3, 1), (3, 2), (3, 3), (3, 4), (3, 5), (3, 6), (4, 1), (4, 2), (4, 3), (4, 4), (4, 5), (4, 6), \\ (5, 1), (5, 2), (5, 3), (5, 4), (5, 5), (5, 6), (6, 1), (6, 2), (6, 3), (6, 4), (6, 5), (6, 6) \end{Bmatrix}.$$

In each pair, the first and second numbers represent the numbers appearing on the first and second die, respectively. Therefore, the discrete random variable $X$ can be considered as follows:

$$\begin{array}{llllll}
(1,1)\to 1, & (1,2)\to 2, & (1,3)\to 3, & (1,4)\to 4, & (1,5)\to 5, & (1,6)\to 6,\\
(2,1)\to 7, & (2,2)\to 8, & (2,3)\to 9, & (2,4)\to 10, & (2,5)\to 11, & (2,6)\to 12,\\
(3,1)\to 13, & (3,2)\to 14, & (3,3)\to 15, & (3,4)\to 16, & (3,5)\to 17, & (3,6)\to 18,\\
(4,1)\to 19, & (4,2)\to 20, & (4,3)\to 21, & (4,4)\to 22, & (4,5)\to 23, & (4,6)\to 24,\\
(5,1)\to 25, & (5,2)\to 26, & (5,3)\to 27, & (5,4)\to 28, & (5,5)\to 29, & (5,5)\to 30,\\
(6,1)\to 31, & (6,2)\to 32, & (6,3)\to 33, & (6,4)\to 34, & (6,5)\to 35, & (6,6)\to 36,
\end{array}$$

***Example*** Think of the hair color of people living in New York City as black, brown, red, and blond. Therefore, the four colors above are members of the sample space, i.e., $\Omega$. We mark these elements as $w_1$, $w_2$, $w_3$ and $w_4$ for black, brown, red, and blond, respectively. We define the random variable $X$ to $w_1$, $w_2$, $w_3$ and $w_4$ to $x_1$, $x_2$, $x_3$, and $x_4$, respectively. That is, $w_1 \to x_1$, $w_2 \to x_2$, $w_3 \to x_3$, and $w_4 \to x_4$. Hence, $x_1, x_2, x_3,$, and $x_4$ vocalize the range of finite discrete random variable $X$.

***Example*** Consider the following sample space of pairs:

$$\Omega = \{(-1,4), (0,4), (1,2), (1,9), (5,-1), (6,3), (7,-2)\}.$$

We consider the random variable $X$ as the distance between two numbers, i.e., $X = \{|i-j| : (i,j) \in \Omega\}$. Thus we have

$$X = \begin{cases} (-1,4) \to 5 \\ (0,4) \to 4, \\ (1,2) \to 1, \\ (1,9) \to 8, \\ (5,-1) \to 6, \\ (6,3) \to 3, \\ (7,-2) \to 9. \end{cases}$$

**Definition** Suppose that $A$ is an event from a sample space $\Omega$.

Suppose that $\Omega$ is the sample space and the event $A$ is from it. Then for each $w \in \Omega$, we define

$$I_A(w) = \begin{cases} 1, & \text{for } w \in A, \\ 0, & \text{for } w \notin A. \end{cases}$$

where $I_A$ is a random variable which indicates an indicator random variable of event $A$.

It is means that, depending on whether event $A$ occurred or not, the values of the random variables $I_A$ are one or zero, respectively.

Thus, we can write

$$\forall w \in \Omega : \begin{cases} I_{\Omega}(w) = 1, \\ I_{\emptyset}(w) = 0. \end{cases}$$

**Definition** If a random variable takes all possible values in its range, it is called continuous random variable.

***Example*** Here is an example of a continuous random variable. Assume that the temperature in Houston in July is between 28 °C and 38 °C in previous years. This means that the temperature can be anywhere between 27.5 °C and 38.5 °C. When the temperature is 30 °C, that is, the temperature is between 29.5 °C and 30.5 °C.

### *2.3.1 Discrete Probability Distribution*

Discrete probability distribution are classified according to the type of random variable support values. If the base of the random variable $X$ is a set of finite or infinite-countable sets, it is called discrete random variable. Of course, it is clear that for any of your finite or infinite sample spaces, the defined random variable must be discrete. But the opposite is not true.

For example, the random variable indicator with support $S = \{0, 1\}$, is defined as follows. Given the sample space $\Omega$ and the event space $A$, the function X is a random variable if written as follows:

$$I_A(w) = 1, \ w \in A, \ I_A(w) = 0, \ w \notin A.$$

**Definition** Suppose that $X$ is a discrete random variable defined on the sample space $\Omega$ with range $R = \{x_1, x_2, ..., x_n\}$. Then, the probability mass function (pmf) of discrete random variable $X$ is expressed as follows:

$$p_x = P(X = x_k), \ k = 1, 2, ..., n,$$

If the following statements holds

I. For every $x \in R, 0 \leq p_x \leq 1$,
II. $\sum\limits_{x \in R} p_x = 1$.

***Example*** Manufacturer's products are classified as superior, average, and bad. In order to represent these states, we propose the random variable $X$, and consider these variables as 5, 3, and 1, respectively. On the other hand, the previous results show a 60, 30, and 10% chance of producing each of these three items, respectively. Therefore,

$$\Omega = \{\text{superior, average and bad}\} \rightarrow^{x} \{1, 3, 5\},$$

we have,

$$X = \begin{cases} bad \to 1, \\ average \to 3, \\ superior \to 5. \end{cases}$$

In this case, the pmf is as follows:

| $x$ | 1 | 3 | 5 |
|---|---|---|---|
| $p_x$ | 0.1 | 0.3 | 0.6 |

That is, the image of $X$ is $\{1, 3, 5\}$.

**Definition** Consider the discrete random variable $X$ with pmf $p_x$. Then for real number $x$ cumulative mass function (cmf) of the discrete random variable $X$, is expressed as follows:

$$F_X(x) = P(X \leq x) = \sum_{k=1}^{x} p_k.$$

According to the probability axioms, for all $x$, $p_x > 0$ and $\sum_k p_x = 1$..

## 2.3.2 *Continuous Probability Distribution*

As mentioned, a discrete random variable is a variable whose values are in a finite set or an infinite but countable sequence. A continuous random variable is a variable for which possible values are in the whole range and do not contain only discrete values. More precisely, a continuous random variable is a variable that has the following two conditions:

1. The range of values may be a range or a social range of several separate ranges. If the possible values of a continuous random variable such as $X$ are an interval, consider both $a$ and $b$ from the set of possible values such that $a < b$, then any other value between $a$ and $b$ is also part of the possible values of the random variable $X$.
2. For any number such as $c$ in the set of possible values of the random variable $X$, the probability that the random variable is exactly equal to $c$ is equal to zero. In other words $P(X = c) = 0$.

For example, we are studying an oil field and sampling it. The depth of wells in this field at different points is a continuous random variable called $X$, which fluctuates

Fig. 2.1 A sample histogram from sample space

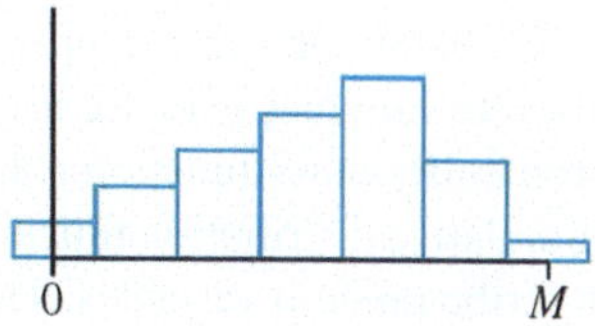


between two numbers $A$ and $B$, so that $A$ is the depth of the shallowest well and $B$ is the depth of the deepest well in this oil field.

Assume that the random variable $X$ is the depth of the oil wells so that $M$ is the maximum depth of the well in meters. As a result, any real number in the interval $[0, M]$ is a possible value for $X$. If we round the values of X to the nearest integer, then the possible values for the well depth of the negative integers are less than or equal to $M$. The discrete distribution of the well depth can be shown with a probabilistic histogram. Assume that the histogram is plotted so that the area of the rectangle above any correct value $k$ is equal to the ratio (percentage) of wells with depth $k$. In this case, the total area of the rectangles will be equal to one (or 100%). One of the sample histograms is shown in Fig. 2.1.

If the depth of the wells is measured more accurately, the histogram indicates the probability of narrowing so that the sum of the area of the rectangles is still equal to one. In this case, the histogram shape has more smoothness than the previous shape (Fig. 2.2).

If we continue in the same way and increase the accuracy of measuring the depth of the wells, eventually the drawing diagram tends towards a continuous curve as shown in Fig. 2.3.

Since the total area of the rectangles in the previous shapes was always equal to one, the area under the curve in the above continuous shape is also equal to one. The probability that the depth of a well is between two numbers $a$ and $b$ is also equal to the area under the curve between two values $a$ and $b$. Figure 2.3 shows the possible continuous distribution.

Fig. 2.2 Two sample histograms from sample space

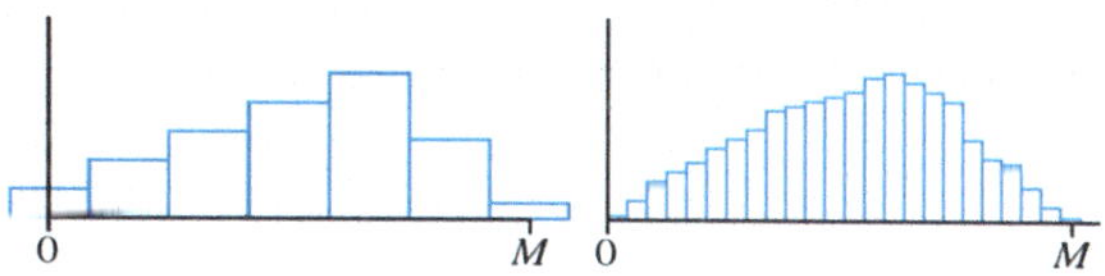


Fig. 2.3 Continuous curve of sample histogram

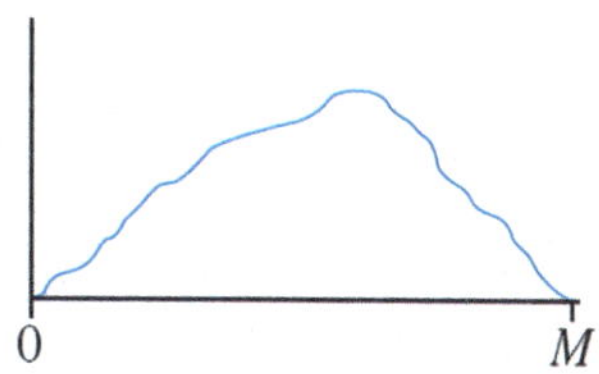

As mentioned in the previous section, one way to describe the distribution of discrete variables is to use the probability mass function (PMF). More precisely, the probability mass function cannot be defined for continuous random variables. The cumulative distribution function (CDF) of a random variable is an alternative way to describe random variables. The advantage of the cumulative distribution function is that it can be defined for any type of random variable. The cumulative distribution function for the random variable $X$ is defined according to the following statement

$$F_X(x) = P(X \leq x), \text{ for all } x \in \mathbb{R}.$$

In order to better understand, let's give an example, which is also presented below.

***Example*** Consider a coin that we toss twice. Suppose $X$ is the number of times you tap. In this case, find the cumulative distribution function $X$. In the first step, we must say that the test is performed twice. In each of these experiments, the probability of being a line is 50%.

In fact, the variable $X$ can have any of the following states ($R_X$ indicates the number of times a tap or line appears):

$$R_X = \{0, 1, 2\}.$$

The probability of each of the above cases is also equal to:

$$P_X(0) = P(X = 0) = \frac{1}{4},$$
$$P_X(1) = P(X = 1) = \frac{1}{2},$$
$$P_X(2) = P(X = 2) = \frac{1}{4}.$$

In order to find the cumulative distribution function (CDF), we do the following. Obviously, the value of $X$ cannot be negative. Therefore, this case can be expressed according to the following statement:

$$F_X(x) = P(X \leq x) = 0, \text{ for all } x < 0.$$

If $0 \leq x < 1$, the probability of occurrence is 0.25. As a result, this situation can be expressed as follows:

$$F_X(x) = P(X \leq x) = P(X = 0), \text{ for } 0 \leq x < 1.$$

The final state is also when x is in the range $1 \leq x < 2$. In this case, the probability of occurrence is equal to:

$$F_X(x) = P(X \leq x) = P(X = 0) + P(X = 1) = \frac{1}{4} + \frac{1}{2} = \frac{3}{4}, \text{ for } 1 \leq x < 2.$$

As a result, we can finally express the probability distribution function of these two launches in the form of the following multidisciplinary function:

$$F_X(x) = \begin{cases} 0 \text{ for } x < 0 \\ \frac{1}{4} \text{ for } 0 \leq x < 1 \\ \frac{3}{4} \text{ for } 1 \leq x < 2 \\ 1 \text{ for } x \geq 2 \end{cases}$$

Note that when you are asked to find the cumulative distribution function of a random variable, you must find the function for all real lines. You should also be careful about discrete random variables when using " > " and " < ". Figure 2.4. represents the function $F_X(x)$.

**Definition** If $X$ is a random variable, the probability distribution function of the random variable $X$ is a function that takes real numbers from 0 to 1. Therefore, we define the probability distribution function as follows:

$$F_X(x) = P(X \leq x),\ x \in \mathbb{R}.$$

Note that $F_X(x)$ is the value of the distribution function of the random variable $X$ at the point $x$. Thus, this function expresses the probability value for the random variable $X$ up to the point $x$. The probability distribution function is also called the cumulative probability function (Fig. 2.5).

Next, we discuss the properties of the random variable probability distribution function.

1. The probability distribution function is a non-descending function, i.e., $x_1 < x_2$ then$F_X(x_1) \leq F_X(x_2)$.
2. The value of the probability distribution function is 1 for the upper bound and zero for the lower bound, that is, $F_X(+\infty) = 1$ and $F_X(-\infty) = 0$.
3. The probability distribution function of a random variable is right-continuous. Thus, if $F_X(x^+)$ is called the right-limit of the probability distribution function at point $x$, then:

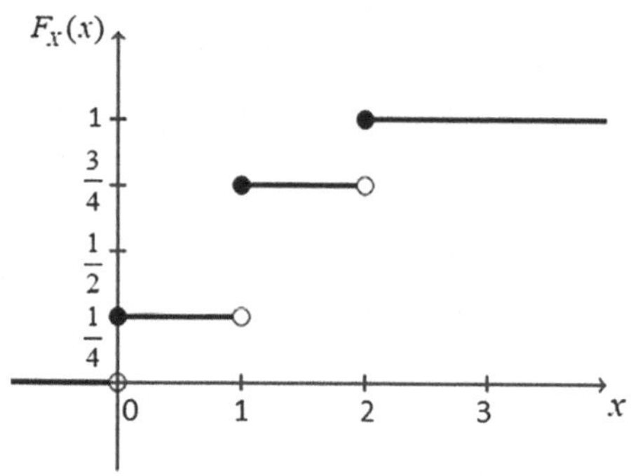


**Fig. 2.4** Cumulative distribution function $F_X(x)$

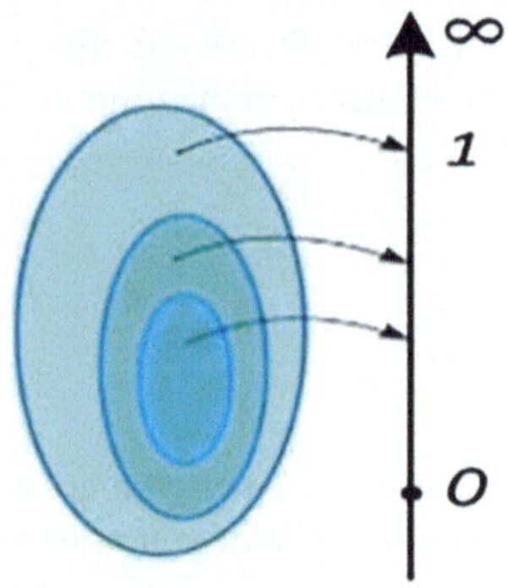


**Fig. 2.5** A sample of diagram of a random variable

$$\lim_{x \to x^+} F_X(x) = F_X(x^+) = F_X(x).$$

4. For the probability distribution function of the random variable $X$ we can write:

$$P(X = x) = F_X(x) = F_X(x^-).$$

5. The following distribution function and rules can also be used to find the probability value:

$$\begin{aligned} P(x < X) &= 1 - F_X(x), \\ P(x_1 < X \leq x_2) &= F_X(x_2) - F_X(x_1), \\ P(x_1 < X \leq x_2) &= F_X(x_2) - F_X(x_1^-), \\ P(x_1 < X \leq x_2) &= F_X(x_2^-) - F_X(x_1^-), \\ P(x_1 < X \leq x_2) &= F_X(x_2^-) - F_X(x_1). \end{aligned}$$

**Definition** Consider the continuous random variable $X$, the probability distribution function (PDF) $f(x)$ such that for both $a$ and $b$, $a$ is less than or equal to $b$, we have:

$$P(a \leq X \leq b) = \int_a^b f(x)dx$$

The above value indicates the probability that $X$ will be in the interval $[a, b]$, which is equal to the area above this interval and below the density function curve, as shown in Fig. 2.6.

The pdf $f(x)$ must be valid under the following two conditions:

1. $f(x) \geq 0$, for all $x$
2. $\int_a^b f(x)dx =$ area under the entire graph of $f(x) = 1$.

Notice that

**Fig. 2.6** Density function curve

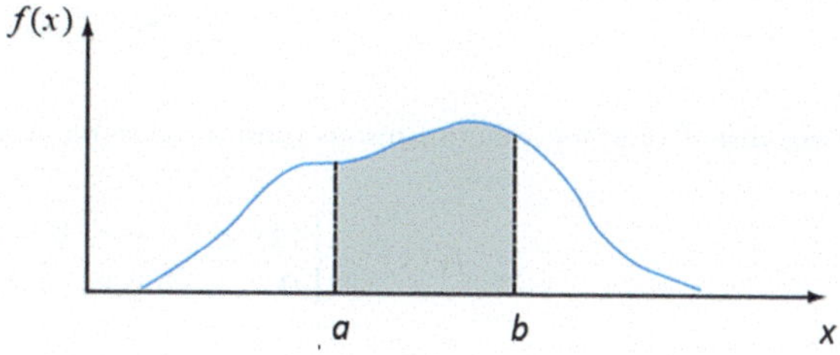


**Fig. 2.7** Integrand of probability density function

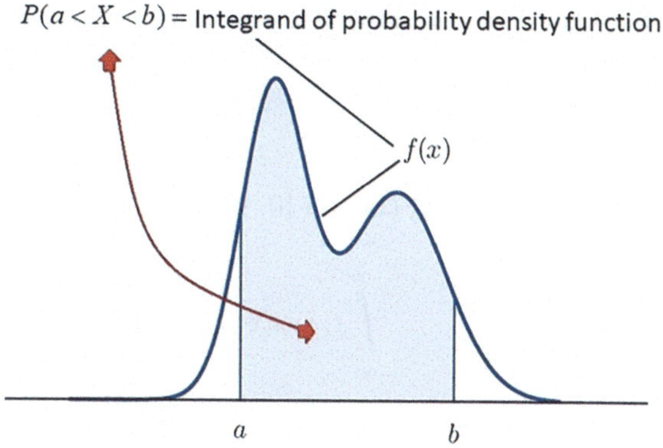


$$f(x) = F_X'(x)$$

Thus, according to the concept of integral, it is clear that this probability is equal to the area below the density function curve in the range $(a, b)$ (Fig. 2.7).

***Example*** Let $X$ be a continuous random variable whose pdf is given by

$$f_X(x) = 2x,\ 0 \le x \le 1.$$

Hence, if we were to find the probability of $X = 0.7$ by simply substituting 0.7 in $f_X(x)$, as we would do in the discrete case, we would have gotten $f_X(x) = 2 \times 0.7 = 1.4$, which is not a probability value since the value is more than 1.

Therefore, if we want to obtain the probability $X = 0.85$ by substituting 0.85 in $f_X(x)$, as we do in the discrete case, we get $f_X(x) = 2 \times 0.85 = 1.7$. Note that it is not a value of probability because its value is greater than 1.

As mentioned, the continuous random variable probability at a point, such as $X = 0.75$, the sample space is zero. Hence, for an example similar to this, we have to ask for the probability of an interval, such [0.65 0.75]. Thus,

$$P(0.65 < x < 0.75) = \int_{0.65}^{0.75} 2x dx$$

$$= 0.75^2 - 0.65^2 = 0.14.$$

***Example*** Consider the continuous random variable $X$ with the following pdf below:

$$f_X(x) = \begin{cases} c(x+2), & 0 \le x \le 10, \\ 0, & \text{o.w.} \end{cases}$$

Find the following statements:

I. A constant $c$ such that $f_X(x)$ is a pdf,
II. $P(X \ge 7)$,
III. $P(2 < X \le 7)$.

Solution

(I) The integral of $f_X(x)$ over [0, 10] is 1. Thus,

$$\int_{-\infty}^{\infty} f_X(x)dx = 1 = \int_0^{10} c(x+2)dx$$

$$= c\left(\frac{x^2}{2} + 2x\right)_0^{10} = 70c = 1.$$

It can be seen that $c = \frac{1}{70}$.
Thus, we obtain

$$f_X(x) = \begin{cases} \frac{1}{70}(x+2), & 0 \le x \le 10, \\ 0, & \text{o.w.} \end{cases}$$

which is surely nonnegative. Hence, $f_X(x)$ defines a pdf.

(II) We have

$$P(X \ge 7) = \int_7^{10} \frac{1}{70}(x+2) = \frac{1}{70}\left(\frac{x^2}{2} + 2x\right)_7^{10} = \frac{63}{140} = 0.45.$$

(III) Moreover, we have

$$P(2 \le X \le 7) = \int_2^{10} \frac{1}{70}(x+2) = \frac{1}{70}\left(\frac{x^2}{2} + 2x\right)_2^{10} = \frac{65}{140} = 0.46.$$

***Example*** We investigate the assets (dollars units) of the occupants of a elderly society in an area. We suppose that the assets value in this state can be defined by a continuous random variable with pdf as follows:

$$f_X(x) = \begin{cases} \frac{2}{x^3}, & x \geq 1 \\ 0, & x < 1 \end{cases}.$$

(a) Prove that $f_X(x)$ is a pdf.
(b) Find the probability that a randomly selected resident asset is between \$300,000 and \$500, 000.
(c) Find the probability that the randomly selected resident asset is at least \$200,000.

Solution

(a) $f_X(x)$ is positive and we have

$$\int_{-\infty}^{\infty} f_X(x)dx = \int_{1}^{\infty} \frac{2}{x^3}dx = -\frac{1}{x^2}\Big|_1^{\infty} = 1.$$

Hence, $f_X(x)$ is a pdf.

(b) $P(3 \leq X \leq 5) = \int_3^5 \frac{2}{x^3}dx = -\frac{1}{x^2}\Big|_3^5 = -\left(\frac{1}{25} - \frac{1}{9}\right) = 0.07.$

(c) $P(X \geq 2) = \int_2^{\infty} \frac{2}{x^3}dx = -\frac{1}{x^2}\Big|_2^{\infty} = -\left(0 - \frac{1}{4}\right) = 0.25.$

***Example*** Assume a random variable such $X$ indicates the lifespan of a lamp in hours. Suppose that the cdf of $X$ is as follows:

$$F_X(x) = \begin{cases} 0, & x < 100, \\ \frac{x^2+x}{990,900}, & 100 \leq x \leq 1000, \\ 1, & x > 1000. \end{cases}$$

(a) Find the pdf.
(b) Find $P(X \geq 800)$.

Solution

(a) Derivativing $F_X(x)$ with respect to $x$ gives

$$f_X(x) = \begin{cases} \frac{2x+1}{990,900}, & 100 \leq x \leq 1000, \\ 0, & x > 1000. \end{cases}$$

(b)

$$\begin{aligned} &P(X \geq 800) = 1 - P(X < 800) = 1 - P(X \leq 800) = 1 - F_X(800) \\ &= 1 - \frac{800^2 + 800}{990,900} = 1 - \frac{640,800}{990,900} = 0.3533. \end{aligned}$$

## 2.4 Joint Distribution

Joint probability distribution or the common probability distribution is discussed in the probability debate, in which the phenomenon in question is interpreted with a set of random variables that are related to it, and the changes of these variables in relation to each other and together (jointly) are examined. In many cases we are interested in studying two or more random variables at the same time. In this connection, for both random variables $X$ and $Y$, we define the cumulative distribution function as follows

$$F_{X,Y}(x, y) = P(X \leq x, Y \leq y), \; -\infty < x, y < \infty.$$

The cumulative distribution function $X$ can be obtained from the joint cumulative distribution function as follows:

$$F_X(x) = P(X \leq x, Y \leq \infty), \; -\infty < x, y < \infty.$$

Thus, we can obtain

$$P(X > x, Y > y) = 1 - F_X(x) - F_Y(b) + F_{X,Y}(x, y).$$

The pdf of random vector $X = (X_1, X_2, \ldots, X_n)$ is as follows:

$$f_{X_1,\ldots,X_n}(x) = \frac{\partial^n}{\partial X_1, \ldots, X_n} F_{X_1,\ldots,X_n}(x_1, \ldots, x_n).$$

Such that

$$f_{X_1,\ldots,X_n}(x) \geq 0,$$

$$\int_{-\infty}^{\infty} \cdots \int_{-\infty}^{\infty} f_{X_1,\ldots,X_n}(x) dx_1 \ldots dx_n = 1,$$

**Definition** For a specified continuous random variable, the pdf of marginal is defined as follows:

$$F_{X_1}(x_1) = \int \cdots \int f_{X_1,\ldots,X_n}(x_1, \ldots, x_n) dx_1 \ldots dx_n.$$

**Definition** Let $X$ be a a random variable, then the pdf of $X$ conditional on random variable $Y$ is defined as:

$$f_{X|Y}(x, y) = \frac{f_{X,Y}(x, y)}{f_Y(y)} \text{ if } f_Y(y) > 0.$$

Therefore, the joint pdf of $X$ can expressed as follows:

$$f_{XY}(x, y) = f_{X|Y}(x, y).f_Y(y),$$

The marginal pdf of continuous random variable $X$ is defined as follows

$$f_X(x) = \int_{-\infty}^{\infty} f_{X,Y}(x, y)dy,$$

Similarly, the marginal pdf of the discrete random variable $X$ is given by

$$f_X(x) = \sum_{-\infty}^{\infty} f_{X,Y}(x, y).$$

***Example*** Assume that $X$ and $Y$ are two discrete random variables, such that.

| | $Y = 1$ | $Y = 2$ | $Y = 3$ |
|---|---|---|---|
| $X = 1$ | $\frac{1}{5}$ | $\frac{1}{7}$ | $\frac{1}{9}$ |
| $X = 1$ | $\frac{1}{10}$ | $\frac{1}{5}$ | $\frac{31}{126}$ |

Find the following:

I. Marginal pmf of $Y$.
II. Probability of $X = 1$ and $Y \geq 2$.
III. Probability of $X = 0$ and $Y < 3$.
IV. Probability of $Y = 2$ conditional on X = 0.

Solution From the table, we have: $S_X = \{0, 1\}$ and $S_Y = \{1, 2, 3\}$.
(I)

$$P_Y(1) = P_{XY}(0, 1) + P_{XY}(1, 1) = \frac{1}{10} + \frac{1}{5} = \frac{3}{10}.$$

$$P_Y(2) = P_{XY}(0, 2) + P_{XY}(1, 2) = \frac{1}{5} + \frac{1}{7} = \frac{12}{35}.$$

$$P_Y(3) = P_{XY}(0, 3) + P_{XY}(1, 3) = \frac{31}{126} + \frac{1}{9} = \frac{5}{14}.$$

Hence,

$$P_Y(y) = \begin{cases} \frac{3}{10} & y = 1, \\ \frac{12}{35} & y = 2, \\ \frac{5}{14} & y = 3, \\ 0 & \text{o.w.} \end{cases}$$

(II)

$$P(X = 1, Y \geq 2) = P_{XY}(1, 2) + P_{XY}(1, 3) = \frac{1}{7} + \frac{1}{9} = \frac{16}{63}.$$

(III)

$$P(X = 0, Y < 3) = P_{XY}(0, 1) + P_{XY}(0, 2) = \frac{1}{10} + \frac{1}{5} = \frac{3}{10}.$$

(IV)

$$P(Y = 2|X = 0) = \frac{P(X = 0, Y = 2)}{P(X = 0)} = \frac{\frac{1}{5}}{\frac{1}{10} + \frac{1}{5} + \frac{31}{126}} = \frac{315}{715}.$$

***Example*** Consider two continuous random variable $X$ and $Y$, with joint pdf

$$f_{XY}(x, y) = e^{-x-y}, \ 0 < x, y < \infty,$$

such that the random variable support is

$$S = \left\{(x, y),\ 0 < y < \frac{x}{7}\right\}.$$

Find the probability that $(X, Y)$ falls in $S$.

Solution

$$\begin{aligned} P\{(X, Y) \in S\} &= \int_0^\infty \int_0^{\frac{x}{7}} e^{-x-y} dxdy \\ &= \int_0^\infty e^{-x}\left[-\left(e^{-\frac{x}{7}} - 1\right)\right]dx = \int_0^\infty e^{-x}\left(1 - e^{-\frac{x}{7}}\right)dx \\ &= \int_0^\infty \left(e^{-x} - e^{-\frac{8x}{7}}\right)dx = \left[e^{-x} + \frac{7}{8}e^{-\frac{8x}{7}}\right]_0^\infty = \frac{1}{8}. \end{aligned}$$

***Example*** Consider two continuous random variable $X$ and $Y$, with joint pdf

$$f_{XY}(x, y) = \frac{2}{3}, \ x^2 \leq y \leq 1, 0 < x < 1.$$

Find

I. The pdf of $Y$ conditional on $X$,

II. The pdf of $Y$, $\frac{1}{9} \le y \le 1$, conditional on $x = \frac{1}{3}$.

Solution

(I) We have

$$f_X(x) = \int_{-\infty}^{\infty} f_{X,Y}(x, y)dy$$

$$= \int_{x^2}^{1} \frac{3}{2}dy = \frac{3}{2}y\Big|_{x^2}^{1} = \frac{3}{2}(1 - x^2),\ 0 < x < 1.$$

Thus,

$$f_{Y|X}(x, y) = \frac{f_{X,Y}(x, y)}{f_X(x)},\ \text{if } f_X(x) > 0$$

$$= \frac{\frac{3}{2}}{\frac{3}{2}(1 - x^2)} = \frac{1}{(1 - x^2)},\ 0 < x < 1,\ x^2 \le y \le 1.$$

(II) For $x = \frac{1}{3}$, we have

$$f_{Y|X}\left(\frac{1}{3}, y\right) = \frac{1}{\left(1 - \frac{1}{9}\right)} = \frac{9}{8},\ x^2 \le y \le 1.$$

## 2.5 Expectation

Expectation is the belief or assumption about what will happen in the future or how someone will behave. It can be based on past experiences, societal norms, or personal desires. Expectations can influence our attitudes, behaviors, and emotions, and can shape our interactions with others. They can also create either positive or negative outcomes depending on whether they are met or unmet. Managing and communicating expectations effectively is important for maintaining healthy relationships, setting realistic goals, and achieving desired outcomes.

In this section, we propose some contents that briefly characterize the behavior of random variables. Focusing the distribution of a random variable around a value is called the mean. The variance shows amount of random variable oscillation around the mean. Using the covariance of two random variables, their behavior can be determined. In other words, covariance expresses whether two random variables deviate from their mean similarly. The covariance matrix for some random vector indicates the variance of each variable and the covariance of each pair. These quantities do not wholly determine the random variable or random vector distribution, however, their behavior can be summarized.

**Definition** Let $X$ be a discrete random variable with range $R$. The expected value of a function, $g(X), g : \mathbb{R} \to \mathbb{R}$ of $X$ is

$$E(g(X)) = \sum_{x \in R} g(x) p_X(x).$$

Similarly, if $X$ and $Y$ are both discrete random variables with ranges $R_X$ and $R_Y$ then the expected value of a function $g(X, Y), g : \mathbb{R}^2 \to \mathbb{R}$ of $X$ and $Y$ is

$$E(g(X, Y)) = \sum_{x \in R_X} \sum_{y \in R_Y} g(x) p_{X,Y}(x, y).$$

If $\vec{X}$ is an $n$-dimensional discrete random vector, the expected value of a function $g(\vec{X}), g : \mathbb{R}^n \to \mathbb{R}$ of $\vec{X}$ is

$$E(g(\vec{X})) = \sum_{x_1} \sum_{x_2} \cdots \sum_{x_n} g(\vec{x}) p_{\vec{X}}(\vec{x}).$$

***Example*** Suppose $X$ is a random variable with probability function $P(X = x)$ as follows. Then the mean of this random variable based on its support is the weighted average based on the weights that indicate the probability of occurrence of each value (Table 2.2).

As can be seen, the value of 1 is more likely to be observed than the others, so it will be more important in calculating the mean. Value 4, on the other hand, is less likely to occur, so it should be less important in calculating the mean. According to the definition of weighted average, the importance of the presence of each value can be considered as the probability of occurrence of each of them. It should be noted that the percentage frequency has the same probability role in the frequency table.

Therefore, the expected value for $X$ can be calculated as follows:

$$1 \times \frac{1}{3} + 2 \times \frac{1}{4} + 3 \times \frac{1}{4} + 4 \times \frac{1}{6} = \frac{8 + 2 \times 6 + 3 \times 6 + 4 \times 4}{24} = \frac{9}{4} = 2.25.$$

Thus, we expect that if we observe the values of the random variable $X$ several times, their average will be 2.25.

**Definition** Let $X$ be a continuous random variable. The expected value of a function $g(X), g : \mathbb{R} \to \mathbb{R}$ of $X$ is

**Table 2.2** A random variable with probability function

| Random variable value of X | 1 | 2 | 3 | 4 |
|---|---|---|---|---|
| Probability event P(X = x) | $\frac{1}{3}$ | $\frac{1}{4}$ | $\frac{1}{4}$ | $\frac{1}{6}$ |

$$E(g(X)) = \int_{x=-\infty}^{\infty} g(x) f_X(x) dx.$$

Similarly, if $X$ and $Y$ are both discrete random variables with ranges $R_X$ and $R_Y$, then the expected value of a function $g(X, Y)$, $g : \mathbb{R}^2 \to \mathbb{R}$ of $X$ and $Y$ is

$$E(g(X, Y)) = \int_{x=-\infty}^{\infty} \int_{y=-\infty}^{\infty} g(x, y) f_{X,Y}(x, y) dx dy.$$

If $\vec{X}$ is an $n$-dimensional discrete random vector, the expected value of a function $g(\vec{X})$, $g : \mathbb{R}^n \to \mathbb{R}$ of $\vec{X}$ is

$$E(g(\vec{X})) = \int_{x_1=-\infty}^{\infty} \int_{x_2=-\infty}^{\infty} \cdots \int_{x_n=-\infty}^{\infty} g(\vec{x}) f_{\vec{X}}(\vec{x}) dx_1 dx_2 ... dx_n.$$

***Theorem*** *For any constant $a \in \mathbb{R}$, any function $g : \mathbb{R} \to \mathbb{R}$ and any continuous or discrete random variable $X$*

$$E(ag(X)) = aE(g(X)).$$

For any constants $a, b \in \mathbb{R}$, any functions $g_1, g_2 : \mathbb{R}^n \to \mathbb{R}$ and any continuous or discrete

random variables $X$ and $Y$

$$E(ag_1(X, Y) + bg_2(X, Y)) = aE(g_1(X, Y)) + bE(g_1(X, Y)).$$

***Proof*** The theorem follows immediately from the linearity of sums and integrals.

**Theorem** *If $X$ and $Y$ are independent random variables defined on the same probability space, and $g, h : \mathbb{R} \to \mathbb{R}$ are univariate real-valued functions, then*

$$E(g(X)h(Y)) = E(g(X))E(h(Y)).$$

***Proof*** We prove the result for continuous random variables, but the proof for discrete random variables is essentially the same.

$$\begin{aligned} E(g(X)h(Y)) &= \int_{x=-\infty}^{\infty} \int_{y=-\infty}^{\infty} g(x)h(y) f_{X,Y}(x, y) dx dy \\ &= \int_{x=-\infty}^{\infty} \int_{y=-\infty}^{\infty} g(x)h(y) f_X(x) f_Y(y) dx dy \end{aligned}$$

$$= E(g(X))E(h(Y)).$$

***Example*** Let the random variables $X$ and $Y$ have a joint pdf defined as

$$f_{X,Y}(x, y) = 2, 0 \leq x \leq y < 1.$$

Find $f_X(x)$, $f_Y(y)$, $E(X)$, $E(Y)$, $Var(Y)$, and $Var(Y)$.

Solution We have

$$f_X(x) = \int_x^1 2dy = 2y|_x^1 = 2(1 - x),\ 0 \leq x < 1,$$

$$f_Y(y) = \int_0^y 2dx = 2x|_0^y = 2y,\ 0 \leq y < 1.$$

Based on the expectation definition, we can write

$$E(X) = \int_0^1 2x(1 - x)dx = 2\left(\frac{x^2}{2} - \frac{x^3}{3}\right)_0^1$$
$$= 2\left(\frac{1}{2} - \frac{1}{3}\right) = \frac{1}{3}.$$

and

$$E(Y) = \int_0^1 2y.ydy = 2\left(\frac{y^3}{3}\right)_0^1 = \frac{2}{3}.$$

Now, for the second moments, we have:

$$E(x^2) = \int_0^1 2x^2(1 - x)dx = 2\left(\frac{x^3}{3} - \frac{x^4}{4}\right)_0^1$$
$$= 2\left(\frac{1}{3} - \frac{1}{4}\right) = \frac{1}{6}.$$

and

$$Var(X) = E(X^2) - (E(X))^2 = \frac{1}{6} - (\frac{1}{3})^2 = \frac{1}{18}$$

and

$$Var(Y) = E(Y^2) - (E(Y))^2 = \frac{1}{2} - (\frac{2}{3})^2 = \frac{1}{18}.$$

One of the interesting features of expectation its monotonicity. Consider two random variables $X$ and $Y$ with $X \leq Y$. Thus, if $Z$ is a random variable derived from the difference between $X$ and $Y$, it is clear that its values will always be negative. As a result, the expectation for such a variable is also negative, and we have:

$$E[Z] = E[X - Y] = E[X] - E[Y] \leq 0 \Rightarrow E[X] \leq E[Y].$$

Moreover, if $X$ and $Y$ are two independent random variables, the following relation can be written for them in terms of expectation:

$$E[XY] = E[X]E[Y].$$

This equation shows that in the case of independence, the average product of two random variables is equal to their average product. Conversely, when the random variables $X$ and $Y$ have a normal distribution and also their product has a normal coefficient of two variables, it can be concluded that they are independent. This means that according to the above conditions, if the expectation of their product is equal to the product of their hopes, the independence of the variables $X$ and $Y$ can be inferred.

If the random variable $X$ has a nonnegative support, we can write the relationship between it and the probability distribution function according to the definition of mathematical hope. Suppose that $F_X(x)$ is a function of the distribution of the random variable $X$ and $E[X]$ is also its mathematical expectation. As a result, the following relationship can be proved between the two for the discrete and continuous states:

$$E[X] = \sum_{i=1}^{\infty} P(X \geq i),$$

$$E[X] = \int_{X \geq 0} (1 - F_X(x))dx.$$

The law of large numbers is one of the most important propositions in the theory of statistics and probability. This law is defined according to the concept of mathematical hope. The law of large numbers states that the mean of random variables is also distributed independently and tends toward their expectation. Suppose that $X_1, X_2, \ldots$ is a sequence of randomly distributed and independent variables with mathematical expectation equal to $\mu$. Then the average mathematical expectation of them, when they are large, is equal to the expectation of each of them. This issue is specified in the following relation:

$$E\left[\frac{X_1 + X_2 + \ldots + X_1}{n}\right] = \mu,\ as\, n \to \infty.$$

This law was last proved by Aleksandr Khinchin, a Russian statistician, in 1929 for any random variable. Of course, you may come across two versions of this law: "Weak Law of Large Numbers" and "Stronge Law of Large Numbers". The first and second laws both specify the same thing, but the space in which the limit is taken is different. Although these rules have two different titles, their main meaning is that as the number of samples increases, the accuracy of estimating the average of the community increases.

### *2.5.1 Mean*

The mean of a random variable is equal to its expected value.

**Definition** The mean or first moment of $X$ is the expected value of $X : E(X)$.

If the distribution tail of a random variable is very heavy, meaning that the probability that a random variable will take large amounts decreases slowly, the average may be infinite. The following example indicates that if the corresponding sum or integral value is not well defined, the mean may not exist.

***Example*** The pdf of the Cauchy random variable is given by

$$f_X(x) = \frac{1}{\pi(1 + x^2)}.$$

By the definition of expected value,

$$E(X) = \int_{-\infty}^{\infty} \frac{x}{\pi(1 + x^2)} dx = \int_{-\infty}^{0} \frac{x}{\pi(1 + x^2)} dx + \int_{0}^{\infty} \frac{x}{\pi(1 + x^2)} dx.$$

Now, by the change of variables $t = x^2$,

$$\int_{0}^{\infty} \frac{x}{\pi(1 + x^2)} dx = \int_{0}^{\infty} \frac{1}{\pi(1 + t)} dt = \lim_{t \to \infty} \frac{\log(1 + t)}{2\pi} = \infty,$$

so $E(X)$ does not exist, as it is the difference of two limits that tend to infinity.

The mean of a random vector is defined as the vector formed by the means of its components.

**Definition** *The mean of a random vector* $\vec{X}$ is

$$E(\vec{X}) = \begin{bmatrix} E(\vec{X}_1) \\ E(\vec{X}_2) \\ \vdots \\ E(\vec{X}_n) \end{bmatrix}.$$

**Theorem** *For any random vector* $\vec{X}$ of dimension $n$, any matrix $A \in \mathbb{R}^{m\times n}$ and $\vec{b} \in \mathbb{R}^m$.

$$E(A\vec{X} + \vec{b}) = AE(\vec{X}) + \vec{b}.$$

***Proof***

$$E(A\vec{X} + \vec{b}) = \begin{bmatrix} E\left(\sum_{i=1}^{n} A_{1i}\vec{X}_i + \vec{b}_1\right) \\ E\left(\sum_{i=1}^{n} A_{2i}\vec{X}_i + \vec{b}_2\right) \\ \vdots \\ E\left(\sum_{i=1}^{n} A_{mi}\vec{X}_i + \vec{b}_m\right) \end{bmatrix} = \begin{bmatrix} \sum_{i=1}^{n} A_{1i}E(\vec{X}_i) + \vec{b}_1 \\ \sum_{i=1}^{n} A_{1i}E(\vec{X}_i) + \vec{b}_1 \\ \vdots \\ \sum_{i=1}^{n} A_{1i}E(\vec{X}_i) + \vec{b}_1 \end{bmatrix} = AE(\vec{X}) + \vec{b}.$$

## 2.5.2 *Median*

Median is a commonly used measure of central tendency in both finance and statistics. In finance, the median is often used to describe the typical value of a set of data without being affected by extreme outliers. For example, the median salary of a group of employees can provide a more accurate representation of the average salary than the mean if there are significant variations in individual salaries. In statistics, the median is used as a measure of central tendency that is more robust to outliers compared to the mean. It is particularly useful when analyzing data that is not normally distributed or when there are extreme values that may skew the results. For example, when analyzing income data, the median income can provide a more accurate representation of the typical income level of a population compared to the mean income. Overall, the median is a valuable tool in finance and statistics for providing a more accurate representation of the central tendency of a set of data, especially in the presence of outliers or non-normally distributed data.

For variables that can be sorted from small to large (significant), the median can be calculated. To get the median, we first sort the data from small to large, then introduce the data that is in the center to the median. The median is an index of which 50% (half) of the observations are smaller and 50% (half) of the observations are larger. One of the disadvantages of median is that it does not consider most information, for example, the median in the set of numbers 110, 90, 24, 20, 73 and the set of numbers

2500, 1900, 24, 0, 73 is the same 73. In fact, the number of numbers is important for calculating the median, and the values at the top and bottom of the median are not important.

**Definition** The median of a discrete random variable $X$ is a number $m$ such that

$$P(X \leq m) \geq \frac{1}{2} \text{ and } P(X \geq m) \geq \frac{1}{2}.$$

The median of a continuous random variable $X$ is a number $m$ such that

$$F_X(m) = \int_{-\infty}^{m} f_X(x)dx = \frac{1}{2}.$$

The following example shows the median strength in the presence of extreme values with non-zero probability.

***Example*** Consider the uniform random variable $X$ with support $[-4.5, 4.5] \cup [99.5, 100.5]$. Then, the mean of $X$ equals

$$\begin{aligned} E(X) &= \int_{x=-4.5}^{4.5} xf_X(x)dx + \int_{x=99.5}^{100.5} xf_X(x)dx \\ &= \frac{1}{10}\frac{100.5^2 - 99.5^2}{2} = 10. \end{aligned}$$

The cdf of $X$ between $-4.5$ and 4.5 is equal to

$$F_X(m) = \int_{-4.5}^{m} f_X(x)dx = \frac{m + 4.5}{10}.$$

We can calculate the median by setting the above equation to 1/2, which is equal to 0.5. Figure 2.8 illustrates the pdf of $X$ and the median and mean location. The median offers a more realistic measurement of the distribution center. The uniform pdf in with mean (red color) 10 amf median (orange color) 0.5 is graphed in Fig. 2.8.

### *2.5.3 Variance and Standard Deviation*

Variance and standard deviation are commonly used in statistics to measure the spread or dispersion of a set of data points. They provide information about how much the data points deviate from the mean of the dataset. Variance is a measure of

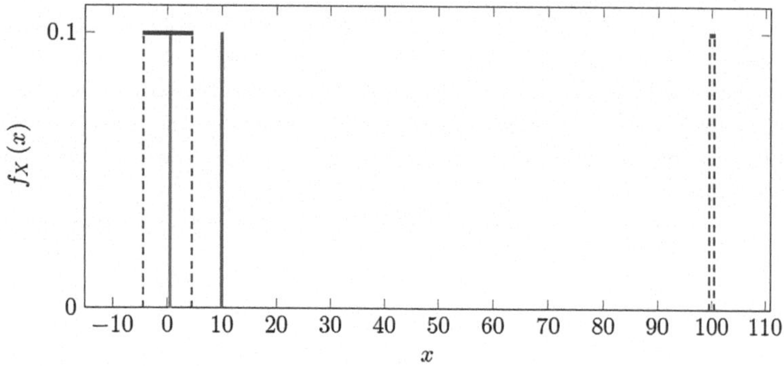


**Fig. 2.8** The uniform pdf in with mean (red color) 10 amf median (orange color) 0.5

how far each data point in a dataset is from the mean. It is calculated by averaging the squared differences between each data point and the mean of the dataset. The variance is useful for understanding the variability or spread of the data points. Standard deviation is the square root of the variance and provides a measure of how spread out the data points are from the mean. It is a commonly used measure of dispersion in statistics and is easier to interpret than the variance because it is in the same units as the data points. Variance and standard deviation are commonly used in descriptive statistics to summarize the spread of data points in a dataset. They provide information about the variability of the dataset and can help identify outliers or extreme values. In inferential statistics, variance and standard deviation are used to estimate the variability of a population based on a sample. They are important for calculating confidence intervals and conducting hypothesis tests. The Variance and standard deviation are used in quality control to monitor and control the variability of a manufacturing process. By calculating these measures, manufacturers can assess the consistency and reliability of their products. The variance and standard deviation are used to measure the risk or volatility of investment returns. Investors use these measures to assess the uncertainty or variability of potential returns and make informed investment decisions. The variance and standard deviation are versatile tools in statistics that provide valuable insights into the spread of data points. They are widely used in various fields, including scientific research, economics, engineering, and healthcare, to analyze and interpret data.

First of all, let us answer the question, what is standard deviation? The standard deviation consists of two words. The first component is the deviation of the distance of each member of a data set from the mean value. The word "standard" also means that this value is standard. The lower the standard deviation of a data set, the closer the data is to the mean and the less scatter. If the standard deviation is a large number, it indicates that the data scatter is high. The standard deviation, then, is a number to indicate the scatter of members of a data set.

**Definition** The mean square or second moment of a random variable $X$ is the expected value of $X^2 : E(X^2)$.

The moments of a probability distribution over a random variable support interval can uniquely describe the properties of that probabilistic distribution. The more complex the distribution, the more moments we need to recognize and retrieve a distribution than other distributions. It is just like in physics: the more complex the behavior of the phenomenon, the more moments we need to describe the behavior of the phenomenon.

If $X$ is a random variable, the moment of the order $p$ ($p$ is a positive integer) is defined as $E[X^p]$.

**Definition** The variance of $X$ is the mean square deviation from the mean

$$\begin{aligned} Var(x) &= E\big((X - E(X))^2\big) \\ &= E(X^2) - E^2(X). \end{aligned}$$

The standard deviation $\sigma_X$ of $X$ is

$$\sigma_X = \sqrt{Var(X)}.$$

**Lemma** *For any constants a and b*

$$Var(aX + b) = a^2 Var(X).$$

***Proof***

$$\begin{aligned} Var(aX + b) &= E\big((aX + b - E(aX + b))^2\big) \\ &= E\big((aX + b - aE(X) - b)^2\big) \\ &= a^2 E\big((X - E(X))^2\big) \\ &= a^2 Var(X). \end{aligned}$$

Inequalities based on probability and used for random variables play an important role in statistics and statistical inference. With the help of these inequalities, we are able to consider boundaries for probabilistic values. Here we will examine two important inequalities called the Markov inequality and the Chebyshev inequality, which play an important role in proving other theorems in statistics, especially mathematical statistics.

Based on Markov inequality, we can find a high limit for the probability value of an event in which a non-negative random variable is greater than a fixed value. This inequality was first introduced by the Russian mathematician Andrey Markov based on the work done by his teacher, Pafnuty Lvovich Chebyshev. This inequality is related to the mathematical probability and expectation of the random variable, and it creates a boundary for the cumulative probability function of the random variable $X$ under certain conditions.

**Theorem** *Let* $X$ be a nonnegative random variable. For any positive constant $a > 0$,

$$P(X \geq a) \leq \frac{E(X)}{a}.$$

***Proof*** Consider the indicator variable $I_{X \geq a}$. We have

$$X - aI_{X \geq a} \geq 0.$$

Since the sum or integral of a non-negative quantity on a real line is positive, the expectation of $X - aI_{X \geq a}$ is non-negative. Due to the fact that expectation is linear and given that $I_{X \geq a}$ is a random variable with expectation $P(X \geq a)$, we have

***Example*** We know the average age of New York University students is 20, but we know of a small number of students over 30. Using Markov's inequality, we want to model the fraction of students over 30 years old with non-negative random variable $A$.

$$P(A \geq 30) \leq \frac{E(A)}{30} = \frac{2}{3}.$$

At most two thirds of the students are over 30.

Given that in some cases little information is available on the random variable distribution, as in the previous example, Markov inequality will not be very useful.

In Chebyshev inequality, the upper bound is determined by the mean and variance of the random variable. Chebyshev inequality is named after its creator, Pafnuty Lvovich Chebyshev. In 1867, the Russian mathematician proved the inequality of his friend and colleague Jules Bienaymé, introduced in 1853. Of course, this inequality was later proved in another way by Markov, one of Chebyshev's students.

**Theorem** *For any positive constant a > 0 and any random variable X with bounded variance*

$$P(|X - E(X)| \geq a) \leq \frac{Var(X)}{a^2}.$$

***Proof*** Using Markov's inequality to the random variable $Y = (X - E(X))^2$ gives the result.

**Corollary** *If* $Var(X) = 0$ *then* $P(X \neq E(X)) = 0$.

***Proof*** Take any $\varepsilon > 0$, by Chebyshev's inequality

$$P(|X - E(X)| \geq \varepsilon) \leq \frac{Var(X)}{\varepsilon^2} = 0.$$

***Example*** You are not very satisfied with your bound on the number of students above 30. You find out that the standard deviation of student age is actually just 3 years. Applying Chebyshev's inequality, this implies that

$$
\begin{aligned}
P(A \geq 30) &\leq P(|A - E(A)| \geq 10) \\
&\leq \frac{Var(A)}{100} = \frac{9}{100}.
\end{aligned}
$$

### 2.5.4 Covariance

Covariance is often used to determine the extent of a linear relationship between two random variables. The value of covariance of the two variables includes positive and negative values. But the closer the value of covariance is to zero, the less linear the dependence between them. Positive values indicate a co-directional relationship between the two variables and negative values of covariance also indicate an inverse relationship between them.

Of course, it should be noted that covariance and how it is calculated (regardless of the sign) does not determine the intensity of the relationship, because this index depends on the unit of measurement of each variable. As a result, the calculation of covariance will be different by changing the unit of measurement of each of the variables. For example, a variable that indicates length may be measured in different units. If the unit of this variable changes from millimeters or centimeters to kilometers, the covariance will decrease, because the numbers related to the calculation of length become smaller, and, conversely, if we specify the length in kilometers instead of millimeters, the covariance will increase.

**Definition** The covariance of $X$ and $Y$ is

$$
\begin{aligned}
Cov(X, Y) &= E((X - E(X))(Y - E(Y))) \\
&= E(XY) - E(X)E(Y).
\end{aligned}
$$

If $Cov(X, Y) = 0$, then $X$ and $Y$ are uncorrelated.

Figure 2.9 illustrates samples of a specified distributions (Gaussian) with various covariances. If the covariance is zero, the joint pdf is spherical. If the covariance is negative, the two variables tend to have similar values with opposite signs. If the covariance is positive, the joint pdf is skewed so that the two variables tend to have similar values. Samples from 2D Gaussian vectors $(X, Y)$, where $X$ and $Y$ are standard Gaussian random variables with zero mean and unit variance, for different values of the covariance between $X$ and $Y$ are presented in Fig. 2.9.

***Theorem***

$$Var(X + Y) = Var(X) + Var(Y) + 2Cov(X, Y).$$

***Proof***

$$Var(X + Y) = E\big((X + Y - E(X + Y))^2\big)$$

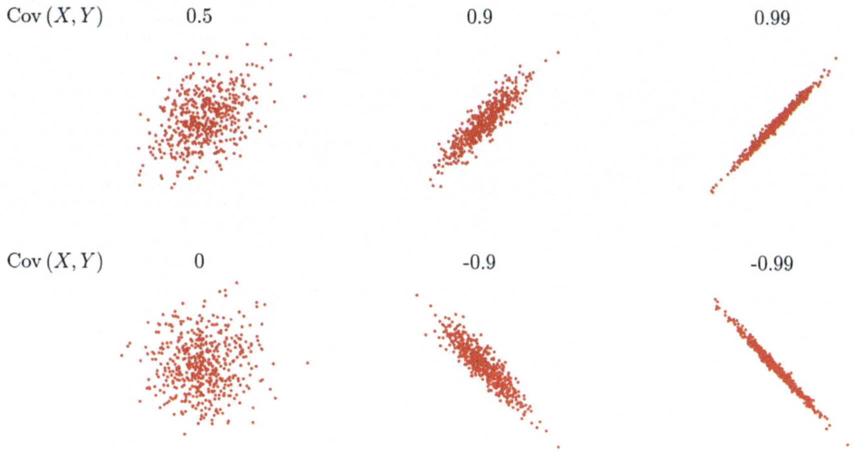


**Fig. 2.9** The behavior of two Gaussian random variables $X$ and $Y$ with different covariances

$$= E\big((X - E(X))^2\big) + E\big((Y - E(Y))^2\big) + 2E((X - E(X))(Y - E(Y)))$$
$$= Var(X) + Var(Y) + 2Cov(X, Y).$$

**Corollary** *If X and Y are uncorrelated, then*

$$Var(X + Y) = Var(X) + Var(Y).$$

The following lemma and example show that independence implies uncorrelation, but uncorrelation does not always imply independence.

**Lemma** *If two random variables are independent, then they are uncorrelated.*

***Proof*** Using definition of covariance and independent random variables for $X$ and $Y$, we have

$$Cov(X, Y) = E(XY) - E(X)E(Y) = E(X)E(Y) - E(X)E(Y) = 0.$$

***Example*** Suppose $X$ and $Y$ are two independent random variables, such that

$$P(X = 0) = P(X = 1) = \frac{1}{2},$$
$$P(Y = 0) = P(Y = 1) = \frac{1}{2}.$$

Define the random variables

$$U = X + Y,$$
$$V = X - Y.$$

Note that

$$\begin{aligned} p_U(0) &= P(X=0, Y=0) = \frac{1}{4}, \\ p_V(0) &= P(X=1, Y=1) + P(X=0, Y=0) = \frac{1}{2}, \\ p_{U,V}(0,0) &= P(X=0, Y=0) = \frac{1}{4} \neq p_U(0)p_V(0) = \frac{1}{8}, \end{aligned}$$

so $U$ and $V$ are not independent. On the other hand, they are uncorrelated as

$$\begin{aligned} Cov(U, V) &= E(UV) - E(U)E(V) \\ &= E((X+Y)(X-Y)) - E(X+Y)E(X-Y) \\ &= E(X^2) - E(Y^2) - E^2(X) + E^2(Y) = 0. \end{aligned}$$

Since $X$ and $Y$ have the same distribution, the final equation holds.

### 2.5.5 Correlation Coefficient

In statistical multivariate analysis, there are different computational methods for measuring the dependence or relationship between two random variables. The correlation coefficient between two variables means the ability to predict the value of one in relation to the other. For example, supply and demand are two interdependent phenomena. One way to show the relationship between two variables is to calculate the covariance or correlation coefficient between them. The larger the value of these two indicators, the greater the relationship or dependence between the two variables. For example, a large correlation can be considered between the two variables of electricity consumption and air temperature. As the temperature rises, so does the use of refrigeration equipment, increasing electricity consumption. As a result, there is a strong correlation between these two variables. Of course, it should be noted that covariance or correlation does not indicate a cause-and-effect relationship, but is only a criterion for showing the degree of dependence between two variables.

**Definition** The Pearson correlation coefficient of two random variables $X$ and $Y$ is

$$\rho_{X,Y} = \frac{E((X - E[X])(Y - E[Y]))}{\sqrt{\text{Var}[X]\,\text{Var}[Y]}}.$$

In this case, the covariance between $X/\sigma_X$ and $Y/\sigma_Y$ is the correlation coefficient between $X$ and $Y$. Figure 2.10 contrasts samples of Gaussian random variables with the identical correlation coefficient and various variances.

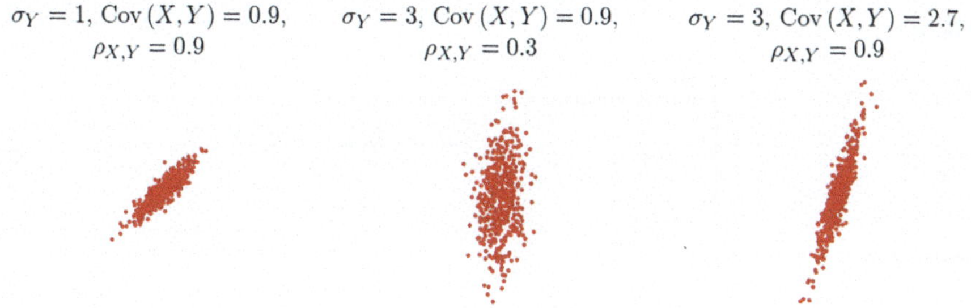


**Fig. 2.10** The behavior of two Gaussian random variables $X$ and $Y$ with identical correlation coefficient and various variances

Although possible, the magnitude of the correlation coefficient is limited by one because the covariance of two random variables cannot exceed the product of their standard deviations. An advantageous explanation of the correlation coefficient is that it specifies the values of $X$ and $Y$ linearly. In fact, if correlation coefficient is 1 or –1, one of the variables is another linear function! These properties are obtained from Cauchy-Schwartz inequality.

**Theorem** (Cauchy–Schwarz inequality)
*For any random variables X and Y defined on the same probability space*

$$|E(XY)| \leq \sqrt{E(X^2)E(Y^2)}.$$

Assume $E(X^2) \neq 0$,

$$E(XY) = \sqrt{E(X^2)E(Y^2)} \Leftrightarrow Y = \sqrt{\frac{E(Y^2)}{E(X^2)}}X,$$

$$E(XY) = -\sqrt{E(X^2)E(Y^2)} \Leftrightarrow Y = -\sqrt{\frac{E(Y^2)}{E(X^2)}}X.$$

**Corollary** *For any random variables X and Y,*

$$Cov(X, Y) \leq \sigma_X \sigma_Y.$$

Equivalently, the Pearson correlation coefficient satisfies

$$|\rho_{X,Y}| \leq 1,$$

with equality if and only if there is a linear relationship between $X$ and $Y$

$$|\rho_{X,Y}| = 1 \Leftrightarrow Y = cX + d,$$

where

$$c := \begin{cases} \dfrac{\sigma_Y}{\sigma_X} \text{ if } \rho_{X,Y} = 1, \\ -\dfrac{\sigma_Y}{\sigma_X} \text{ if } \rho_{X,Y} = -1, \end{cases} \qquad d := E(Y) - cE(X).$$

***Proof*** Let

$$U := X - E(X),$$
$$V := Y - E(Y).$$

From the definition of the variance and the correlation coefficient,

$$E(U^2) = Var(X)$$
$$E(V^2) = Var(Y)$$
$$\rho_{X,Y} = \frac{E(UV)}{\sqrt{E(U^2)E(V^2)}}$$

The result now follows from applying Cauchy–Schwarz inequality to $U$ and $V$.

The covariance matrix of a random vector represents the interaction between the components of the vector. This includes the variance of every component in diagonal and the covariance between various components in off-diagonals.

**Definition** The covariance matrix of a random vector $\vec{X}$ is defined as

$$\sum_{\vec{X}} = \begin{bmatrix} Var(\vec{X}_1) & Cov(\vec{X}_1, \vec{X}_2) & \dots & Cov(\vec{X}_1, \vec{X}_n) \\ Cov(\vec{X}_2, \vec{X}_1) & Var(\vec{X}_2) & \dots & Cov(\vec{X}_2, \vec{X}_n) \\ \vdots & \vdots & \ddots & \vdots \\ Cov(\vec{X}_n, \vec{X}_1) & Cov(\vec{X}_n, \vec{X}_2) & \dots & Var(\vec{X}_n) \end{bmatrix}$$
$$= E(\vec{X}\vec{X}^T) - E(\vec{X})E(\vec{X})^T.$$

If all inputs of a vector are uncorrelated, its covariance matrix is diagonal. The following theorem states an expression for the linear conversion covariance matrix of a random vector.

**Theorem** *Let $\vec{X}$ be a random vector of dimension n with covariance matrix $\sum$. For any matrix $A \in \mathbb{R}^{m\times n}$ and $b \in \mathbb{R}^m$,*

$$\sum_{A\vec{X}+b} = A \sum_{\vec{X}} A^T.$$

***Proof***

$$
\begin{aligned}
\sum_{A\vec{X}+b} &= E\left((A\vec{X}+\vec{b})(A\vec{X}+\vec{b})^T\right) - E(A\vec{X}+\vec{b})E(A\vec{X}+\vec{b})^T \\
&= AE(\vec{X}\vec{X}^T)A^T + \vec{b}E(\vec{X})A^T + AE(\vec{X})\vec{b} + \vec{b}\vec{b}^T \\
&- AE(\vec{X})E(\vec{X})^T A^T - AE(\vec{X})\vec{b}^T - \vec{b}E(\vec{X})^T A^T - \vec{b}\vec{b}^T \\
&= A\left(E(\vec{X}\vec{X}^T) - E(\vec{X})E(\vec{X})^T\right)A^T \\
&= A\sum_{\vec{X}} A^T.
\end{aligned}
$$

### 2.5.6 Conditional Expectation

In probability theory, the conditional expectation of a random variable is its expected value, given that there is a specific set of conditions. If a random variable can take only a limited number of values, the condition is that the variable can only take a subset of those values. Depending on the context, the conditional expectation can be a random variable or a function. The random variable is similarly represented by the conditional probability $E[X|Y]$. The form of the function is either $E[X|Y = y]$ or the symbol of a separate function such as $f(y)$ with the meaning $E[X|Y] = f(y)$ is introduced.

$$E(g(X, Y)|X = x) = \sum_{y \in R} g(x, y) f_{Y|X}(y|x) dy,$$

if $Y$ is discrete and has range $R$, whereas

$$E(g(X, Y)|X = x) = \int_{y=-\infty}^{\infty} g(x, y) f_{Y|X}(y|x) dy,$$

if $Y$ is continuous.

Note that $E(g(X, Y)|X = x)$ can actually be interpreted as a function of $x$, since it maps every value of $x$ to a real number. This allows us to define the conditional expectation of $g(X, Y)$ given $X$ as follows.

**Definition** The conditional expectation of $g(X, Y)$ given $X$ is

$$E(g(X, Y)|X) := h(X),$$

where

$$h(X) := E(g(X, Y)|X = x).$$

**Theorem** (Iterated expectation)
*For any random variables X and Y and any function* $g : \mathbb{R}^2 \to \mathbb{R}$

$$E(g(X, Y)) = E(E(g(X, Y)|X)).$$

***Proof*** We demonstrate the consequence for continuous random variables, the evidence for discrete random variables and for values that depend on continuous and discrete random variables is almost the same. For this purpose, we set

$$\begin{aligned} h(x) :&= E(g(X, Y)|X = x) \\ &= \int_{y=-\infty}^{\infty} g(x, y) f_{Y|X}(y|x)dy. \end{aligned}$$

Now,

$$\begin{aligned} E(E(g(X, Y)|X)) &= E(h(X)) \\ &= \int_{x=-\infty}^{\infty} h(x) f_X(x)dx \\ &= \int_{x=-\infty}^{\infty} \int_{y=-\infty}^{\infty} f_X(x) f_{Y|X}(y|x) g(x, y)dydx \\ &= E(g(X, Y)). \end{aligned}$$

***Example*** Let us compute the mean time at which the car breaks down, i.e. the mean of $T$. By iterated expectation

$$\begin{aligned} E(T) &= E(E(T|M, R)) \\ &= E\left(\frac{1}{M + R}\right) \\ &= \int_0^1 \int_0^1 \frac{1}{m + r} dmdr \\ &= \int_0^1 \log(r + 1) - \log(r)dr \\ &= \log 4 \approx 1.39. \end{aligned}$$

***Example*** Let us compute the mean weight of a bear in Yosemite National Park. By iterated expectation

$$E(W) = E(E(W|S))$$
$$= \frac{E(W|S=0) + E(W|S=1)}{2} = 180kg.$$

***Example*** Let us compute the mean of the coin-flip outcome $X$. By iterated expectation

$$E(X) = E(E(X|B))$$
$$= E(B) = \int_0^1 2b^2 db = \frac{2}{3}.$$

### 2.5.7 Moment Generating Function

Moment in statistics is a quantitative measure or indicator to determine the shape of a probability function or the distribution of a random variable. This index is used in mathematics, especially statistics, as well as in mechanics. If this function is written based on the physical properties of an object, then the zero moment represents the same mass. Also, the first moment divided by the total mass expresses the center of gravity. On the other hand, the second moment will be proportional to the inertia of an object.

In this way, the moment *s* in physics reflect the properties of the body. This is also the case in statistics, which means that with the help of moment *s*, the properties of random variables are examined. The concept and use of moment in statistics will also be determined as the following: The zero moment for a random variable is equal to the total probability or value of 1. The first moment represents the mean and the second central moment also determines the variance. The third and fourth moment will also be proportional to the skewness and kurtosis of the random variable probability distribution.

The concept of moment in statistics is very close to physics. This means that to recognize a random variable (or in physics to identify an object), its moments must be identified. The moment s of a probability distribution over a random variable support interval can uniquely describe the properties of that probabilistic distribution. The more complex the distribution, the more moment we need to recognize and retrieve a distribution than other distributions. It is just like physics that the more complex the behavior of the body, the more moment it takes to describe the object behavior.

If present, the moment s can vary from zero to infinite moment, each of which is used to represent the property or characteristic of a random variable.

The $n$th moment of a real function such as $f(x)$ is defined in terms of the variable $x$ around the center $c$ as follows:

$$\mu_n = \int_{-\infty}^{\infty} (x - c)^n f(x)dx.$$

Note that often in calculations, the moment of a function is considered when $c$ is equal to zero and that moment is called *raw moment* and denoted by $\mu'$. If the value of $c$ is the average of the values of $x$, it is called the *central moment.*

Moments can have values other than natural numbers. For example, the inverse moment of the first order is written as follows:

$$\mu_{(-n)} = \int_{-\infty}^{\infty} (x - c)^{-n} f(x)dx.$$

They often use central moment instead of zero moment, because we can get more information from such moments.

Just as we defined moment s for a function in terms of its variables, we can also define moments for a function and its random variable, such as $X$. If $X$ is a random variable with a distribution function $F_X(x)$, then the $n$th-order moment for the random variable $X$ is displayed and calculated as follows:

$$\mu_n' = E[X] = \int_{-\infty}^{\infty} x^n f(x)dx.$$

Note that the variable is random $X$ in this definition may not have a probability density function and is calculated based on the Riemann–Stieltjes cumulative distribution and integral function. It is clear that the expected value capabilities of a random variable are represented as zero central moments.

If for a random variable such as $X$ the following integral does not exist (it is infinite), then we say that its $n$th order moment does not exist

$$E[|X^n|] = \int_{-\infty}^{\infty} |x^n| dF_X(x) = \infty,$$

As the name implies, the moment generating function can be used to generate moment s. The moment generating function for the random variable $X$ will be written as follows:

$$M_X(t) := E[e^{tX}] = \int_{-\infty}^{\infty} e^{tx} dF_X(x),\ t \in \mathbb{R}.$$

As long as this expectation is finite, the moment generating function can be used to generate the distribution moment s of a random variable.

***Note:*** It is clear that if the random variable $X$ has a discrete distribution, the sum will be used to calculate the expectation in equation above and the integral if it is continuous.

$$M_X(t) = \sum_{i=1}^{\infty} e^{tx_i} p_i,$$

$$M_X(t) = \int_{-\infty}^{\infty} e^{tx} f(x) dx,$$

provided that the probability density function is present in the continuous state.

According to the MacLaurin series of exponential functions, we can write

$$e^{tX} = 1 + tX + \frac{t^2X^2}{2!} + \frac{t^3X^3}{3!} + \ldots + \frac{t^nX^n}{n!} + \ldots$$

As a result, the moment generating function for the random variable $X$ will be a polynomial in terms of $t$ as follows:

$$\begin{aligned} M_X(t) = E[e^{tX}] = 1 + tE[X] + \frac{t^2E[X^2]}{2!} + \frac{t^3E[X^3]}{3!} + \ldots + \frac{t^nE[X^n]}{n!} + \ldots \\ = 1 + tM_1 + \frac{t^2m_2}{2!} + \frac{t^3m_3}{3!} + \ldots + \frac{t^nm_n}{n!} + \ldots \end{aligned}$$

In the above relation, $m$ represents the *ith* moment around zero.

## 2.6 Important Discrete Random Variables

In this section we describe several discrete random variables that are useful for probabilistic modeling.

### *2.6.1 Discrete Uniform*

In probability theory, distributions of random variables are of great importance for recognizing random phenomena. Among these, discrete uniform distribution and continuous uniform distribution have many applications, especially in simulating data related to other statistical distributions.

**Definition** Let $X$ be a random variable. We divide the $[a, b]$ into $k$ subintervals, each of size $(b-a)/k$ where $k$ is a natural number and $a \leq k \leq b$. In other words, the endpoints of the subintervals are $a, a+k, a+2k, ..., b$. Therefore, the probability, $p_x$, of $X$, is defined as:

$$p_X = P(X = x) = \begin{cases} \dfrac{k}{b-a+k}, & a \leq k \leq b, x = a, a+k, a+2k, ...b, \\ 0, & \text{o.w.} \end{cases}$$

It is called the discrete uniform pmf. $X$ is referred to as the uniformly distributed random variable.

Suppose that $a = 1,\ k = 1,$ and $b = n$. In this case, we have

$$P_X = P(X = x) = \begin{cases} \dfrac{1}{n}, & x = 1, 2, 3, ...n, \\ 0, & \text{o.w.} \end{cases}$$

It is clear that since $1/n > 0$, for each $x = 1, 2, 3, ..., n,$ and $\sum_{k=1}^{n} \frac{1}{n} = 1$, and $p_X$ for special case is a pmf. Moreover, we can obtain the expectation and variance of discrete uniform random variable $X$

$$E(X) = \frac{(n+1)}{2},\ Var(X) = \frac{(n^2-1)}{12},\ M(t) = \left(\frac{e^t}{n}\right)\left(\frac{1-e^{nt}}{1-e^t}\right).$$

***Example*** Consider a random variable $X$ that represents the appropriate dice roll test. In this case, $X$ will be a uniform discrete random variable. Figure 2.11 illustrates pmf of uniform discrete random variable $X$.

### *2.6.2 Bernoulli*

A random test that has only two values and the results of which are predictable is a Bernoulli test. Bernoulli test results are usually characterized by success or failure. For example, tossing a coin to see head is a Bernoulli experiment. Because if we consider success as a tail observation and failure as a tail observation, the results of this randomized experiment will have only two values of success or failure. One of the most important conditions for a Bernoulli test is that the probability of success or failure is constant. This means that by repeating this experiment under the same conditions, the probability of success does not change. It should be noted that these repetitions should be independent of each other. The name Bernoulli was chosen for this type of experiment based on the name of the Swiss mathematician Jacob Bernoulli, who did a lot of research in this field.

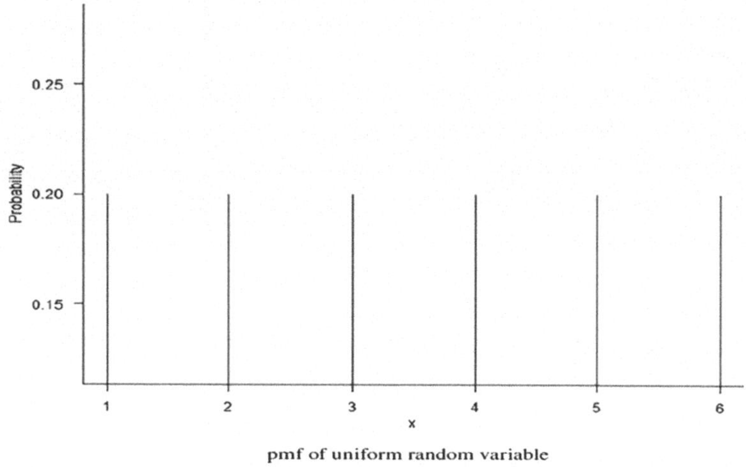


**Fig. 2.11** The pmf of uniform discrete random variable $X$

**Definition** *(Bernoulli)*

The pmf of a Bernoulli random variable with parameter $p \in [0, 1]$ is given by

$$p_X(0) = 1 - p,$$
$$p_X(1) = p.$$

Letting $q = 1 - p$, the pmf Bernoulli random variable can be rewritten as:

$$p_x = P(X = x) = \begin{cases} p^x q^{1-x}, & x = 0, 1, \\ 0, & \text{o.w.} \end{cases}$$

Moreover, the cmf of Bernoulli random variable is expressed:

$$cmf(X) = \begin{cases} 0, & x < 0, \\ q, & 0 \leq x < 1, \\ 1, & x \geq 1. \end{cases}$$

To obtain the mean, variance, and moment generating function of Bernoulli random variable $X$, we provide Table 2.3.

Therefore,

$$E(X) = p,\ Var(X) = pq,\ M(t) = q + pe^t.$$

***Example*** Assume $X$ is a random variable that represents a defective coin toss test so that the probability of showing a head is $p = 2/3$ and a tail is $q = 1/3$. In this case,

**Table 2.3** The mean, variance, and moment generating function of Bernoulli random variable

| X | 1 | 0 | Sum |
|---|---|---|---|
| $pX(x)$ | $p$ | $q$ | $1$ |
| $xpX(x)$ | $p$ | 0 | $p$ |
| $X^2$ | 1 | 0 | |
| $X^2pX(x)$ | $p$ | 0 | $p$ |
| $e^{tx}pX(x)$ | $pe^t$ | $q$ | $q+pe^t$ |
| $Z^xpx(x)$ | $zp$ | $q$ | $Q+pz$ |

we have $p + q = 2/3 + 1/3 = 1$. If 0 and 1 represent tail and head, respectively, $X$ has values of 0 and 1. Therefore, $P(X = 0) = 1/3$ and $P(X = 1) = 2/3$. Thus, we can write

$$p_x = P(X = x) = \begin{cases} \frac{1}{3}, & x = 0, \\ \frac{2}{3}, & \text{x} = 1. \end{cases}$$

or

$$p_x = P(X = x) = \left(\frac{2}{3}\right)^x \left(\frac{1}{3}\right)^{1-x}, \ x = 0, 1.$$

Whenever a Bernoulli experiment is repeated $n$ times with the same probability of success and failure independently in every experiment, the sample space have $2^n$ elements.

***Example*** The toss of a coin will be the result of each head (H) or tail (T) test. So, for a throw, we have $\Omega = \{H, T\}$ that contains 21 points. For 2 coin tosses, the sample space $\Omega = \{(T, T), (H, H), (T, H), (H, T)\}$ that contains the $4 = 2^2$ elements. That is, the first and second throws come out H, the first one comes out H and the second T, the first comes out T and the second H, and both come out T (Table 2.4).

Also, the sample space for three coin tosses, will have the following $8 = 2^3$ elements:

$$\Omega = \{(T, T, T), (H, H, H), (H, T, H), (T, H, T), (H, H, T), (T, H, H), (T, T, H), (H, T, T)\}.$$

The probabilities of the $\Omega$ elements are triple as mentioned in the Table 2.5.

**Table 2.4** The sample space and event probability of 2 coin tosses

| 1 | 2 | 3 | 4 | Sum |
|---|---|---|---|---|
| (H,H) | (H,T) | (T,H) | (T,T) | $(p+q)^2=1$ |
| $p^2$ | pq | pq | $q^2$ | |

**Table 2.5** The sample space and event probability of 3 coin tosses

| 1 | 2 | 3 | 4 | 5 | 6 | 7 | 8 | Sum |
|---|---|---|---|---|---|---|---|---|
| HHH | HHT | HTH | THH | HTT | THT | TTH | TTT | 1 |
| $p^3$ | $p^2q$ | $p^2q$ | $p^2q$ | $pq^2$ | $pq^2$ | $pq^2$ | $q^3$ | |

Now we assume that the coin is biased with the probability that one head will occur in the form of an $5/8$, and will be dealt three times independently. Then, the head is likely to appear only in the second experiment

$$P(T, H, T) = pq^2 = \left(\frac{5}{8}\right)\left(\frac{3}{8}\right)^2 = \frac{45}{125} = 36\%.$$

Likewise, the three heads probability is $(\frac{5}{8})^3 = 24\%$, while that of having three tails in a row is $\left(\frac{3}{8}\right)^3 = 5\%$.

## 2.6.3 Binomial

In some cases, the number of successes or failures in $n$ tests may be important to us rather than the order in which the results occur in repeated independent tests.

**Definition** Suppose the random variable $X$ shows the number of successes in $n$ repeated independent experiments with success probability of $p$ and failure of $q = 1 - p$. This random variable is called a binomial random variable. A binary variable is a variable that has two possible outcomes. For instance, gender (male/female) or passing a test (yes/no) are both examples of a binary random variable.

The pmf of a binomial random variable $X$ for $j$ successes is defined as $B_j(j; n, p)$ and we have

$$P(X = j) = B_j(j; n, p) = \binom{n}{j} p^j q^{n-j},\ j = 0, 1, 2, ..., n,\ q = 1 - p.$$

Generally, pmf of a binomial random variable, i.e., $B_j(j; n, p)$, characterizes the treatment of a random variable such as $X$ with the following statements:

I. The number of experiments is constant.
II. Every experiment is independent of the next.
III. Every experiment has two results.
IV. Probability of success in any experiment is $p$.

It is clear that the cmf for binomial random variable $X$ can be defined as follows:

$$P(X \leq j) = \sum_{k=0}^{[j]} \binom{n}{k} p^k q^{n-k},$$

***Example*** Assume the grade point average of a particular academy is graded on a 4-point scale. Also suppose that the grade point average of 20% of students is more than 3.5. From all the students of this academy, ten students are randomly selected. The random variable $X$ shows the total number of students out of *10* with a grade point average of more than *3.5*. Then $X$ is a binomial random variable and its pmf diagram is illustrated in Fig. 2.12.

***Example*** 30 people are enrolled in a driving instruction class. The chance of passing the course for a person who passes the final exam is equal to 70%. The instructor wants to know the successful chances of 20 out of 30 people in the driving test. To do this, he can use a binomial distribution that expresses the number of people accepted out of 20. To find the desired probability, with $p = 0.7, j = 20$ and $n = 30,$ we have

$$B_j(20; 30, 0.7) = \binom{30}{20} (0.7)^{20} (0.3)^{10} = 0.1416 \approx 14\%.$$

***Example*** A family has 5 children, including both boys and girls. If the probability of being a boy or a girl is equal, we obtain the pmf of boys and girls in the family.

Solution Suppose $X$ indicates the number of daughters in the family. We consider each child to be a test with a probability of success (daughter) of $p$ and failure (boy) of $1 - p$. Suppose consecutive births are independent and the probability for daughters

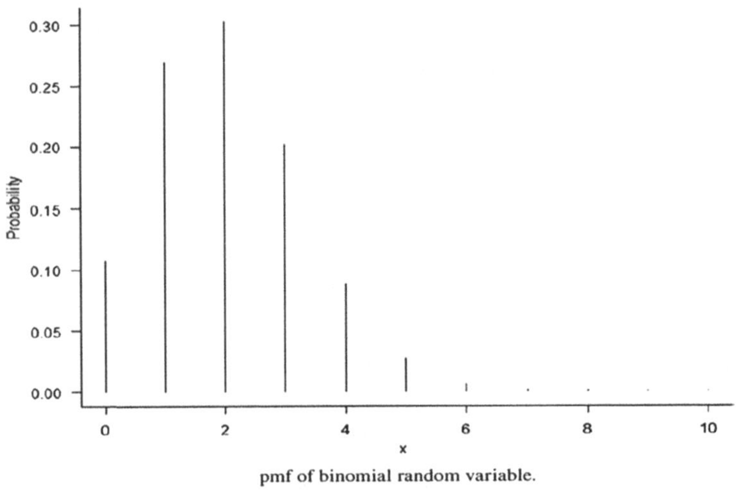


**Fig. 2.12** The pmf of binomial random variable

and boys is 1/2. Therefore, we have:

$$P(\text{exactly } x \text{ girls}) = \text{P}(\text{X} = \text{x}) = \binom{5}{x}\left(\frac{1}{2}\right)^x\left(\frac{1}{2}\right)^{5-x}$$
$$= \binom{5}{x}\left(\frac{1}{2}\right)^5.$$

We have Table 2.6.

**Theorem** *For a binomial random variable X with $B_j(j; n, p)$, $n = 0, 1, \ldots$, letting $q = 1 - p$, we have the following properties:*

$$E(X) = np,$$
$$Var(X) = np(1-p) = npq,$$
$$M(t) = \left(pe^t + (1-p)\right)^n = (pe^t + q)^n.$$

***Proof*** We have

$$M(t) = E(e^{tx}) = \sum_{j=0}^{n} e^{tj} p_j = \sum_{j=0}^{n} e^{tj} \binom{n}{j} p^j q^{n-j}$$
$$= \sum_{j=0}^{n} e^{tj} \binom{n}{j} (pe^t)^j (1-p)^{n-j} = \left(pe^t + (1-p)\right)^n.$$

$$E(X) = \sum_{j=0}^{n} j \binom{n}{j} p^j q^{n-j} = M'(0) = n(p + 1 - p)p = np.$$

$$E(X^2) = \sum_{j=0}^{n} j^2 \binom{n}{j} p^j q^{n-j} = M''(0)$$
$$= n(n-1)\left(pe^t + (1-p)\right)^{n-2}(pe^t)^2 + n\left(pe^t + (1-p)\right)^{n-1}\Big|_{t=0}$$
$$= n(n-1)p^2 + np.$$
$$Var(X) = n(n-1)p^2 + np - (np)^2 = np(1-p).$$

**Table 2.6** A random variable with pmf

| X | 0 | 1 | 2 | 3 | 4 | 5 |
|---|---|---|---|---|---|---|
| pmf, $p_X(x)$ | 1/32 | 5/32 | 10/32 | 10/32 | 5/32 | 1/32 |

### 2.6.4 Geometric

Consider a series of independent Bernoulli randomized trials in which the probability of success for each is constant and equal to *p*. If we consider the random variable *X* as the number of experiments to achieve the first success, this random variable will have a probability distribution with a geometric name. Thus, there is a relationship between Bernoulli's experiment and the random variable and geometric distribution.

For example, suppose a doctor in a village examines people for the first sign of diabetes. The probability of the onset of diabetes is predetermined and equal to *p*. The chance that a doctor will examine the symptoms of diabetes by examining the 5th person can be calculated from the geometric distribution. In this scenario, it is possible to compute the probability of seeing the first success in the first throw, the second throw, and so forth.. So the set of values or the basis of this random variable will be $x = 1, 2, ....$ As a result, this random variable is similar to Bernoulli's random variable and two discrete sentences, except that the statistical population described by the geometric random variable is infinite.

As mentioned, the geometric random variable is the number of repetitions of an experiment until the first success in a sequence of Bernoulli random experiments. To further emphasize and clarify the characteristics that should repeat the Bernoulli experiment, we make the following points:

- The sequence of randomized trials must be independent of each other.
- The sequence of randomized trials should be Bernoulli.
- The parameters of the Bernoulli test sequence must be constant.

According to the definition of this random variable, the form of the probability function or its probability mass is as follows:

$$P(X = x) = p(1 - p)^{x-1} = pq^{x-1},\ x = 1, 2, ...$$

Of course, sometimes the definition of a geometric random variable may be thought of as the number of fractures required to achieve the first success. In this case, if *Y* is considered the number of failures required to achieve the first success, the fulcrum *Y* is equal to $y = 0, 1, 2, 3, ...$ and its probability function is as follows:

$$P(Y = y) = p(1 - p)^{y-1} = pq^{y},\ \mathrm{y} = 0, 1, 2, ...$$

Thus, the relationship $X = Y + 1$ is established between two random variables *X* and *Y*.

Suppose we throw a fair coin until we see the first head. If the number of throws is considered as a random variable, the probability function is in the form of a geometric distribution with parameter $\frac{1}{2}$, and the probability that we observe a head in the second throw is calculated as follows:

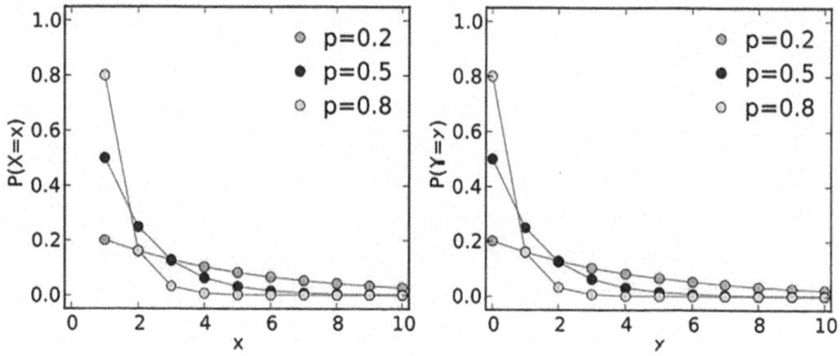


**Fig. 2.13** Probability function of geometric random variable for $P(X = x) = p(1-p)^{x-1}$ and $P(Y = y) = p(1-p)^{y-1}$ with different $x$ and probability of success $p$

$$P(X = x) = \frac{1}{2}\left(1 - \frac{1}{2}\right)^{2-1} = 0.25.$$

If we draw the probability value for the support of a random variable for such a case, a graph like Fig. 2.13 is created, which for the probability of success $p = 0.5$ points are marked in purple. This graph is similar to an exponential function with a base smaller than one.

In Fig. 2.13, the value of the random variable is plotted on the horizontal axis and the probability value is plotted on the vertical axis. The points are also connected using a line to show the shape of the geometric distribution curve. In this figure, according to the support of the random variable, the curve related to the probability of a geometric random variable is displayed in the form $Y$ (number of fractures), and the probability function for the geometric random variable is displayed in the form $X$ (number of tests).

***Example*** Assume 20% of collegians in a academy have a grade point average above 3.5. You choose collegians at random until you meet the first collegian with a grade point average of at least 3.5. Suppose that $X$ is a random variable that represents the total number of collegians until their grade point average is 3.5 or higher. In this case, the random variable $X$ according to the Fig. 2.14 represents the geometric random variable.

The geometric random variable cmf is expressed as follows.

$$\sum_{x=1}^{\infty} p_x = \sum_{x=1}^{\infty} p(1-p)^{x-1} = 1 - (1-p)^n, \ \text{n} = 1,2,\ldots,$$

and

$$\sum_{x=0}^{\infty} p_x = \sum_{x=0}^{\infty} p(1-p)^{x} = 1 - (1-p)^{n+1}, \ \text{n} = 0,1,2,\ldots,$$

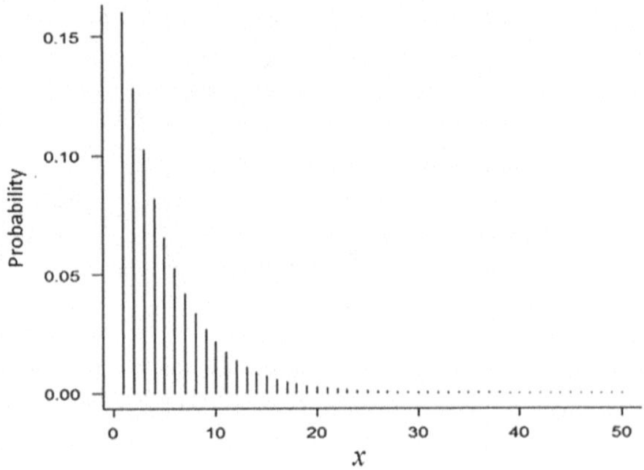


**Fig. 2.14** The pmf of geometrric random variable

Thus,

$$\sum_{x=1}^{\infty} p_x = \sum_{x=1}^{\infty} p(1-p)^{x-1}$$
$$= p\sum_{x=0}^{\infty}(1-p)^x.$$

***Example*** How many times must a die be rolled for the number 3 to appear for the first time?

Solution We put the event of "the number of other numbers until the number 3 to appear for the first time by a random variable $X$".

Consider a 6-sided die where appearance of the six numbers has the same probability as $1/6$. In this case, $X$ follows a geometric distribution as follows:

$$p_n = P(X = n) = \left(\frac{1}{6}\right)\left(\frac{1}{6}\right)^n, \; n = 0, 1, 2, \ldots$$

After $n$ rolls, the probability that the number 3 will appear for the first time is

$$p_0 = P(X = 0) = \left(\frac{1}{6}\right)\left(\frac{5}{6}\right)^0 = \left(\frac{1}{6}\right) \approx 17\%.$$
$$p_1 = P(X = 1) = \left(\frac{1}{6}\right)\left(\frac{5}{6}\right)^1 = \left(\frac{5}{36}\right) \approx 14\%.$$
$$p_2 = P(X = 2) = \left(\frac{1}{6}\right)\left(\frac{5}{6}\right)^2 = \left(\frac{25}{2166}\right) \approx 12\%.$$
$$p_3 = P(X = 3) = \left(\frac{1}{6}\right)\left(\frac{5}{6}\right)^3 = \left(\frac{125}{1296}\right) \approx 10\%.$$

$$p_4 = P(X = 4) = \left(\frac{1}{6}\right)\left(\frac{5}{6}\right)^4 = \left(\frac{625}{7776}\right) \approx 8\%.$$

$$p_5 = P(X = 5) = \left(\frac{1}{6}\right)\left(\frac{5}{6}\right)^5 = \left(\frac{3125}{46,656}\right) \approx 7\%.$$

$$p_6 = P(X = 6) = \left(\frac{1}{6}\right)\left(\frac{5}{6}\right)^6 = \left(\frac{15625}{279,936}\right) \approx 6\%.$$

$$p_7 = P(X = 6) = \left(\frac{1}{6}\right)\left(\frac{5}{6}\right)^7 = \left(\frac{78125}{1679616}\right) \approx 5\%.$$

As $n$ approaches infinity, $\left(\frac{5}{6}\right)^n$ approaches 0, and hence, $p_\infty = 0$. Of course,

$$\sum_{n=0}^{\infty}\left(\frac{1}{6}\right)\left(\frac{5}{6}\right)^n = \left(\frac{1}{6}\right)\sum_{n=0}^{\infty}\left(\frac{5}{6}\right) = \left(\frac{1}{6}\right)\left(\frac{1}{1-\frac{5}{6}}\right) = 1.$$

**Theorem** *The expected value of a geometric random variable X in Case 1 and Case 2, respectively, is as follows:*

$$E(X) = \frac{1}{p},$$

and

$$E(X) = \frac{1-p}{p},\ x = 0, 1, 2, \ldots$$

***Proof*** Case 1:

$$\begin{aligned} E(X) &= \sum_{x=1}^{\infty} x p_x = \sum_{x=1}^{\infty} x p(1-p)^{x-1} \\ &= p\left(\sum_{x=1}^{\infty}(1-p)^{x-1} + \sum_{x=2}^{\infty}(1-p)^{x-1} + \sum_{x=3}^{\infty}(1-p)^{x-1} + \cdots\right) \\ &= p\left(\frac{1}{p} + \frac{1-p}{p} + \left(\frac{1-p}{p}\right)^2 + \cdots\right) \cdots = 1 \\ &+ (1-p) + (1-p)^2 + \cdots = \frac{1}{p}. \end{aligned}$$

Case 2:

$$
\begin{aligned}
E(X) &= \sum_{x=0}^{\infty} x p_x = \sum_{x=0}^{\infty} x p(1-p)^x \\
&= p \sum_{x=0}^{\infty} (1-p)^{x-1} = p(1-p) \sum_{x=0}^{\infty} x(1-p)^{x-1}
\end{aligned}
$$

but

$$
\sum_{x=0}^{\infty} x(1-p)^{x-1} = \frac{d}{dp}\left(\sum_{x=0}^{\infty} (1-p)^x\right).
$$

The series on the right hand side represents a geometric series that converges to $1-p < 1$. Assuming the convergence of the series, the summation and differentiation can be interchanged.

Now using the infinite geometric progression, we have:

$$
\begin{aligned}
\sum_{x=0}^{\infty} x(1-p)^{x-1} &= -\frac{d}{dp}\left(\frac{1}{1-(1-p)}\right) \\
&= \frac{1}{p^2}.
\end{aligned}
$$

Thus,

$$
\begin{aligned}
p(1-p) \sum_{x=0}^{\infty} x(1-p)^{x-1} &= \frac{p(1-p)}{p^2} \\
&= \frac{1-p}{p}.
\end{aligned}
$$

***Example*** What is the expectation value of times the number 2 first appears on a dice roll?

Solution We have

$$
\begin{aligned}
p_X(x) &= \left(\frac{1}{6}\right) = \left(\frac{5}{6}\right)^{x-1}, \ x = 1, 2, 3 \\
E(X) &= \frac{1-p}{p} = \frac{\frac{5}{6}}{\frac{1}{6}} = 5.
\end{aligned}
$$

**Theorem** *For geometric random variable X, we have*

Case 1:

$$
M(t) = \frac{pe^t}{1-(1-p)e^t}, \ 0 < p < 1.
$$

$$Var(X) = \frac{1-p}{p^2}.$$

Case 2:

$$M(t) = \frac{p}{1-(1-p)e^t}, \; 0 < p \le 1.$$
$$Var(X) = \frac{1-p}{p^2}.$$

***Proof*** We will prove case 2 and leave the other proof to the reader as an exercise. we have:

$$M(t) = E(e^{tX}) = \sum_{x=0}^{\infty} P(X = x)e^{xt}$$

Then,

$$\sum_{x=0}^{\infty} P(X = x)e^{xt} = \sum_{x=0}^{\infty} (1-p)^x p e^{xt}$$
$$= p \sum_{x=0}^{\infty} \left((1-p)e^t\right)^x$$

We deal with sum of infinite geometric sequence, for this sum to be convergent we must have:

$$\left|(1-p)e^t\right| < 1.$$

In the case $p = 1$, this demand is satisfied immediately regardless of $t$. Otherwise, as both $e^t$ and $1-p$ are positive:

$$e^t < \frac{1}{1-p}.$$

Logarithm of power, implies that

$$t < -\ln(1-p)$$

is the range of $t$ for which the expectation is well-defined.

Now applying sum of infinite geometric sequence, we have for this range of $t$ :

$$M(t) = p \sum_{x=0}^{\infty} \left((1-p)e^t\right)^x = \frac{p}{1-(1-p)e^t}.$$

***Example*** Referring to previous example, we have:

$$Var(X) = \frac{1-\frac{1}{6}}{\left(\frac{1}{6}\right)^2} = 30.$$

Memoryless is a very important feature of geometric distribution.

**Theorem** *Suppose that X is a geometric random. In this case*

$$P(X \geq r + x | X \geq r) = P(X \geq r).$$

where r and s are two integers.

***Proof*** We have:

$$\begin{aligned} P(X \geq x) &= 1 - P(X < x) = P(X \leq x - 1) \\ &= 1 - \left(1 - (1-p)^{(x-1)+1}\right) = (1-p)^x. \end{aligned}$$

Therefore,

$$P(X \geq r + x | X \geq r) = P(X \geq x) = (1-p)^x.$$

Thus,

$$\begin{aligned} P(X \geq r + x | X \geq r) &= \frac{P(X \geq r + x)}{P(X \geq r)} \\ &= \frac{(1-p)^{r+x}}{(1-p)^r} \\ &= (1-p)^x. \end{aligned}$$

## 2.6.5 *Negative Binomial*

In probability and statistics theory, the negative binomial distribution is known as a discrete distribution. Given the relationship between this distribution and the Bernoulli experiment, it is clear that many random phenomena may follow this distribution. The distribution of negative binomial is used especially in medical and industrial studies. If the random variable $X$ is considered as the number of successes in a sequence of independent Bernoulli experiments and the distribution before $r$ failure, the probability distribution of $X$ can be considered as negative binomial. For example, the number of days a device can cause a problem follows the distribution of negative binomial, provided that the probability of it working every day is constant.

As another example, the number of shots on goal until the $r$th of the goal can be considered as a random variable of negative binomial. Here it can be assumed that achieving a goal is a defeat and not scoring a goal is a success.

Suppose we have a sequence of Bernoulli random experiments with a chance of success of $p$. In this case, the result of each experiment has a probability of $p$ equal to *1* and a probability of $1-p$ equal to *0*. We know that a value of *1* indicates success and *0* indicates failure. This test continues until $r$ fails. If $X$ is the number of successes by the end of the experiment, we can say that the distribution of this random variable is negative binomial with parameters $r$ and $p$. In this case we write $X \sim NB(x; r, p)$ and read $X$ has a negative binomial distribution with parameters $r$ and $p$. According to the definition of this random variable, it is clear that its basis is all non-negative integers.

**Definition** Suppose $X$ is a discrete random variable. In this case, we consider its probability distribution as negative binomial if its pmf is as follows:

$$NB(x; r, p) = \binom{x + r - 1}{x} p^r (1 - p)^x, \ r > 0, \ x = 0, 1, \ldots$$

We leave it as an exercise to show that the pmf of negative binomial can be rewritten as:

$$NB(x; r, p) = \binom{-r}{x} p^r (1 - p)^x, \ r > 0, \ x = 0, 1, \ldots$$

To show that $NB(x; r, p)$, indeed, is a pmf, we note $NB(x; r, p)$ is positive and

$$\begin{aligned}\sum_{x=0}^{\infty} NB(x; r, p) &= \sum_{x=0}^{\infty} \binom{x + r - 1}{x} p^r (1 - p)^x \\ &= \sum_{x=0}^{\infty} (-1)^x \binom{-r}{x} p^r (1 - p)^x \\ &= (1 - p)^{-r} (1 - p)^r = 1.\end{aligned}$$

***Example*** A company conducts agricultural studies for groundwater. According to studies, drilling in a well has a 25% chance of producing water. What is the probability that the first strike was drilled in the fourth well?

Solution To solve this problem, we use a geometric distribution and simulate the problem with a geometric random variable. Hence, by setting $p = 0.25, 1 - p = 0.75, n = 4,$ and $x = 1$,as $r = n - x = 4 - 1 = 3$ we have:

$$P(X = 3) = \binom{3 + 1 - 1}{3} p^1 (1 - p)^{4-1} = \binom{2}{3} (0.25)(0.75)^3 = 0.1055 \approx 11\%.$$

Given the geometric distribution and assuming that $X$ represents the number of experiments before the first success (strike), we will have:

$$p(x = 4) = (0.25)(0.75)^3 = 0.1055 \approx 11\%.$$

**Selling candy problem**

A student needs money for a summer tour. To make money, he decides to sell some candy to his neighbors. There are 37 houses in his neighborhood. If he sells 5 candies, he has provided the cost of the scientific tour.

What is the probability that he will sell the last (5th) candy in the $n$th house? Of course, we also have the condition that the probability of buying candy in each house is fixed and equal to 0.6.

Recall that one of the modes of calculating the probability function for a random variable of two negative experiments is the probability $k$ of failure to achieve the $r$ of success in $r+k = n$ Bernoulli experiment with the success probability $p$ (parameter), provided that the result of the last experiment is success. In this case, selling 5 candies corresponds to 5 successes. The number of tests, which is equal to $n = k + 5$, also indicates the number of houses that the student must visit for sale.

Thus, we can write

$$X \sim NB(x; 5, 0.4).$$

Thus, based on the probability function of a random variable, we have two negative binomial

$$P(X = n) = \binom{(n-5)+5-1}{n-5}(1-0.)^5 0.4^{n-5} = \binom{n-1}{n-5} 2^5 \frac{3^{n-5}}{5^n}.$$

Of course, it should be noted that $n \geq 5$ is considered here. Thus, the probability that he sold 5 candies after passing the tenth house is equal to *0.010033*. Also, the probability that 5 candies were sold before reaching the ninth house is calculated as follows:

$$P(X = 5) + P(X = 6) + P(X = 7) + P(X = 8) = 0.01024 + 0.03072 + 0.055298 + 0.0774144 = 0.17367.$$

**Theorem** *Let X be a negative binomial random variable. Then, expectation, variance, and moment generating function of X is expressed as follows:*

$$E(X) = \frac{pr}{(1-p)}, \; Var(X) = \frac{pr}{(1-p)^2}, \; M(t) = \left(\frac{1-p}{1-pe^t}\right)^r, \; t < \ln p.$$

Proof of this is left to the reader as an exercise.

### 2.6.6 Hypergeometric

The random variable we are to describe is very much similar to the binomial random variable we discussed earlier. A sample is a portion, subset, or a part of the population of interest (finite or infinite number of them). A sample selected such that each element or unit in the population has the same chance to be selected is called a random sample. Sampling from a population is sometimes referred to as random draw.

Now, suppose we have a finite population of size $N$, $N = 0, 1, 2, \ldots$. We draw a sample of size $n$ from this population without replacement. There are two types of outcomes from each draw in this experiment: success and failure. Thus, each sample consists of two subsamples, one the success observations and the other failure observations, with total number of outcomes as $n$, which is the total number of draws, where $n = 0, 1, 2, \ldots, N$.

Based on the description of the scenario, the probability of success and failure changes on each draw due to the change in the remaining population as a draw is made. This feature is the difference between the binomial pmf and hypergeometric pmf.

Now, we let $S$ denote the total number of successes in the population, that is, $S = 0, 1, 2, \ldots, N$.

Of interest is calculating the probability of having $x$, $\max(0, n + S - N) \leq x \leq \min(n, S)$, successes in the sample.

To respond, we first argue as follows. Let us denote the random variable $X$ to represent the number of successes. Looking for $x$ successes would mean $n - x$ failures. The number of ways $x$ successes can occur from the total successes in the population, $S$, is $\binom{S}{x}$ and of failures is $\binom{N-S}{n-x}$. Occurrence of the successes and failures is the product of the two, that is, $\binom{S}{x}\binom{N-S}{n-x}$.It is assumed the choosing $n$ object from a total of $N$ objects is equiprobable. Thus, we can now state the following definition.

**Definition** Let $X$ represent the number $x$ of type 1 outcomes with pmf defined by:

$$p_x = P(X = x) = \frac{\binom{S}{x}\binom{N-S}{n-x}}{\binom{N}{n}}, \quad x = \max[0, n + S - N, \ldots, \min(n, S)].$$

Then, $X$ is called the hypergeometric random variable and equation above is referred to as the hypergeometric pmf.

Let us list the symbols involved in the definition above.

- $N$: Population.
- $n$: Total number of draws.
- $S$: The number of successes included in the population.
- $x$: Number of observed successes.

$p_X(x)$ is a pmf because it is obviously positive when $\max(0, n + S - N) \leq \min(n, S)$, and applying geometric series sum, we have:

$$\sum_{0 \leq x \leq n} \frac{\binom{S}{x}\binom{N-S}{n-x}}{\binom{N}{n}} = 1.$$

***Example*** Consider a junior statistics class with 30 students enrolled. 20 of the students are mathematics majors and 10 other science majors. To assess some measures regarding students' learning, we draw 10 students without replacement. What is the probability that there are exactly six mathematics majors in the drawing?

Solution In this drawing, we consider the inclusion of a mathematics major as a success. We now summarize the problem's information in Table 2.7.

**Table 2.7** A summary of information about a class of 20 math students and 10 other science students

| $N = 30$ | $X = 6$ | $S = 20$ | $n = 10$ |
|---|---|---|---|
| | Drawn | Not Drawn | Total |
| Mathematics majors | $x = 6$ | $S - x = 20 - 6 = 14$ | $S = 20$ |
| None mathematics major | $n - x = 4$ | $N + x - n - S = 6$ | $N - S = 10$ |
| Total | $n = 10$ | $N - n = 30$ | $N = 30$ |

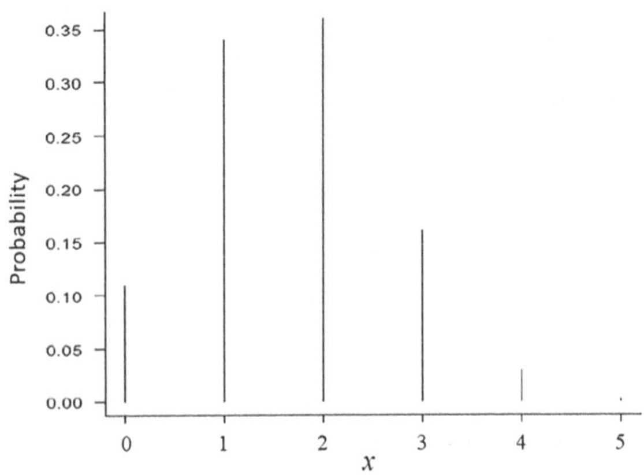


**Fig. 2.15** The pmf of Hypergeometric random variable

Thus, we have

$$p_x(6) = P(X = 6) = \frac{\binom{20}{6}\binom{30-20}{10-6}}{\binom{30}{10}} = \frac{\binom{20}{6}\binom{10}{4}}{\binom{30}{10}} = \frac{\frac{20!}{6!4!}\frac{10!}{4!6!}}{\frac{30!}{10!20!}} = \frac{8,139,600}{30,045,015} \approx 27\%.$$

***Example*** Consider a box with 30 light bulbs and assume that it contains 20 nondefective and 10 defective bulbs. Suppose a quality controller randomly selects five bulbs without replacements. Let X be the random variable that represents the number of defective bulbs he finds in the selected five. Then, the random variable X follows a hypergeometric distribution, and the pmf of X can be graphed as in (Fig. 2.15).

### *2.6.7 Poisson*

We will discuss now the final standard discrete random variable. A Poisson pmf assists in describing the chances of occurrences of a number of events in some given time interval or in a given space when the value of average number of occurrences of the events is known. Consider, for instance, a random experiment where we observe the number of occurrences of independent events during a given time interval such that two or more events do not occur at the same time, and the events are occurring at the same rate.

In many standard queues (waiting lines), the arrival to a waiting room (buffer) is assumed to follow Poisson pmf. Here are other examples of number of events occurring in an interval that Poisson pmf can be applied: telephone calls ringing in an office per hour, the number of cars arriving at a traffic light every 3 min, the number of patients arriving at an emergency room every hour, the number of mutations on a given strand of DNA over a unit time, the number of particles emitted by a radioactive source in a given time, and the number of computer network breakdowns over a 24-hour period.

The Poisson random variable and pmf is named after the French mathematician Siméon Denis Poisson (1781–1840).

**Definition** A nonnegative random variable $X$, with a one parameter $\lambda$ and pmf

$$p_j = P(X = j) = \frac{e^{-\lambda}\lambda^j}{j!}, \ j = 0, 1, \ldots,$$

is called the Poisson random variable. The parameter $\lambda$ is the shape parameter, which indicates the average number of events in the given time interval. The base $e$ is Euler's constant, which is approximately 2.71828.

It can easily be seen that $p_j \geq 0$. We also see that

$$\sum_{j=0}^{\infty} p_j = \sum_{j=0}^{\infty} \frac{e^{-\lambda}\lambda^j}{j!} = e^{-\lambda} \sum_{j=0}^{\infty} \frac{\lambda^j}{j!} = e^{-\lambda} e^{\lambda} = 1.$$

Thus, $p_j$, indeed, is a pmf. The cdf for Poisson pmf is

$$P(X \le x) = \sum_{j=0}^{x} \frac{e^{-\lambda}\lambda^j}{j!}, \; j = 0, 1, \ldots$$

***Example*** Suppose that computers in a computer laboratory at a university break on an average of 1.5 computers per day. What is the probability of breakdown of

I. Three computers in a particular day?
II. No more than four breakdowns in a day?
III. At least five breakdowns on a particular day?

Solution Let $X$ be a random variable representing the number of breakdown of computers in the laboratory per day. We assume that breakdown of computers follows a Poisson pmf with the average rate of $\lambda = 1.5$.

Now, for part (I), we have:

$$p_3 = P(X = 3) = \frac{e^{-1.5}(1.5)^3}{3!} = 0.1255 \approx 13\%.$$

For part (II), we are to find

$$P(0 \le X \le 4) = P(X = 0) + P(X = 1) + P(X = 2) + P(X = 3) + P(X = 4).$$

$$p_0 = P(X = 0) = \frac{e^{-1.5}(1.5)^0}{0!} = 0.2231 \approx 22\%.$$

$$p_1 = P(X = 1) = \frac{e^{-1.5}(1.5)^1}{1!} = 0.3347 \approx 33\%.$$

$$p_2 = P(X = 2) = \frac{e^{-1.5}(1.5)^2}{2!} = 0.2510 \approx 25\%.$$

$$p_3 = P(X = 3) = \frac{e^{-1.5}(1.5)^3}{3!} = 0.1255 \approx 13\%.$$

$$p_4 = P(X = 4) = \frac{e^{-1.5}(1.5)^4}{4!} = 0.0471 \approx 5\%.$$

Thus,

$$P(0 \le X \le 4) = 0.2231 + 0.3347 + 0.2510 + 0.1255 + 0.0747 = 0.9814 \approx 98\%.$$

For part (III), we are to find

$$P(X \ge 5) = P(X = 5) + P(X = 6) + \cdots$$

$$p_0 = P(X = 5) = \frac{e^{-1.5}(1.5)^5}{5!} = 0.0141 \approx 1\%$$

$$p_6 = P(X = 6) = \frac{e^{-1.5}(1.5)^6}{6!} = 0.0035 \approx 0\%.$$

Therefore,

$$P(X \geq 5) = 0.0141 + 0.0035 + \ldots = 0.0176 + \ldots$$

However, to obtain a precise value, we can use:

$$\begin{aligned} &P(X \geq 5) = 1 - P(X < 5) = 1 - P(X \leq 4) \\ &= 1 - (P(X = 0) + P(X = 1) + P(X = 2) + P(X = 3) + P(X = 4)) \\ &= 1 - 0.9814 = 0.0186 \approx 2\%. \end{aligned}$$

Of course, as expected, from parts (II) and (III), the sum of probabilities is 1.

***Example*** The number of meteorites seen in the night sky can be modeled with a Poisson distribution. In the following, we will deal with the conditions of Poisson distribution and the number of visible meteorites.

- If $X$ is the number of times a meteorite is seen in a given period of time, the basis of this random variable is natural numbers.
- The number of meteors seen at each time interval is independent of the next.
- The average number of meteors visible at regular intervals is assumed. This rate is approximately equal to 5 meteors per hour.
- The probability of seeing two meteorites at the same time is very low and the simultaneous occurrence of these events can be ignored.

According to the explanations given, it is clear that the parameter for the Poisson distribution in this example is equal to 5. Because the average number of meteors seen per unit time (1 h) is 5.

In this way we can calculate the probability value of λ, which represents the observation of 5 meteorites per hour

$$P(X = 5) = e - \lambda \frac{\lambda^x}{x!} = e^{-5} \frac{5^5}{5!} = 0.1755.$$

In order to better observe this distribution, we have separately plotted the possible distribution for observing meteorites in the diagram below. It is quite clear that based on this, the probability of seeing 5 meteors per hour is higher than other values (Fig. 2.16).

When $np$ is kept constant, a significant relation among Poisson and binomial distributions is that the binomial distribution function approaches the Poisson distribution function.

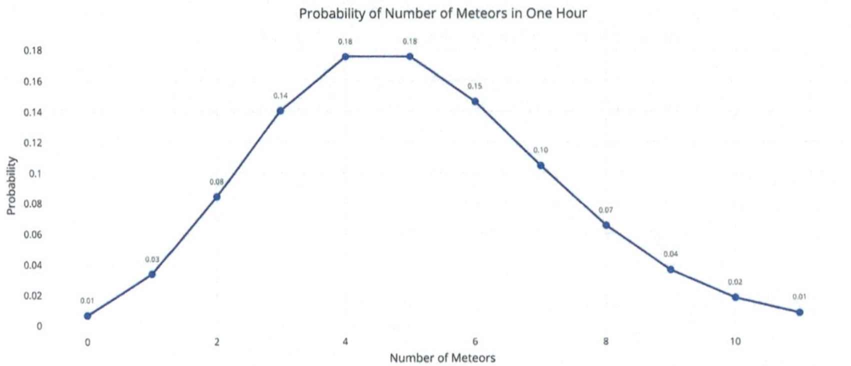


**Fig. 2.16** Probability of number of meteors in one hour that the average number of meteors is approximately equal to 5 meteors per hour

**Theorem** *Let X be a Poisson random variable with parameter λ. Then,*

$$E(X) = Var(X) = \lambda,\ M(t) = e^{\lambda(e^t - 1)}.$$

***Proof*** From Poisson pmf, the *rth* moment of the Poisson random variable *X* is

$$E(X^r) = \sum_{x=0}^{\infty} x^r \frac{e^{-\lambda}\lambda^x}{x!}.$$

Therefore, for $r = 1$ we obtain

***Example*** In a car factory, the probability that the car will be returned due to a technical violation in the quality control section is 1%. What is the probability that 5 out of 300 production devices will be returned?

$$E(X) = \sum_{x=0}^{\infty} x \frac{e^{-\lambda}\lambda^x}{x!} = \lambda e^{-\lambda} \sum_{x=0}^{\infty} x \frac{\lambda^{x-1}}{x(x-1)!}$$
$$= \lambda e^{-\lambda}\left(1 + \lambda + \frac{\lambda^2}{2!} + \frac{\lambda^3}{3!} + \ldots\right) = \lambda e^{-\lambda} e^{\lambda} = \lambda.$$

For $r = 2$, we have

$$E(X^2) = \sum_{x=0}^{\infty} x^2 \frac{e^{-\lambda}\lambda^x}{x!} = \sum_{x=0}^{\infty}\left([x(x-1) + x]\frac{\lambda^x e^{-\lambda}}{x!}\right)$$
$$= \sum_{x=0}^{\infty}\left(x(x-1)\frac{e^{-\lambda}\lambda^x}{x!}\right) + \sum_{x=0}^{\infty}\left(x\frac{e^{-\lambda}\lambda^x}{x!}\right)$$

$$
\begin{aligned}
&= \sum_{x=0}^{\infty}\left(x(x-1)\frac{e^{-\lambda}\lambda^x}{x(x-1)(x-2)!}\right) + \sum_{x=0}^{\infty}\frac{e^{-\lambda}\lambda^x}{(x-1)!}\\
&= \sum_{x=0}^{\infty}\frac{e^{-\lambda}\lambda^x}{(x-2)!} + \sum_{x=0}^{\infty}\frac{e^{-\lambda}\lambda^x}{(x-1)!}\\
&= \lambda^2 e^{-\lambda}\sum_{x=0}^{\infty}\frac{\lambda^{x-2}}{(x-2)!} + E(X)\\
&= \lambda^2 e^{-\lambda}\left[1+\lambda+\frac{\lambda^2}{2!}+\frac{\lambda^3}{3!}+\ldots\right]+\lambda\\
&= \lambda^2 e^{-\lambda}e^{\lambda}+\lambda = \lambda^2+\lambda.
\end{aligned}
$$

Hence,

$$Var(X) = E(X^2) - E(X)^2 = \lambda^2 + \lambda - \lambda^2 = \lambda.$$

Solution Here the average number of returns is the Poisson distribution parameter. That is $\lambda = 300 \times 0.01 = 3$.

Thus, to calculate the probability, we have

$$P(X = 5) = \frac{e^{-3}3^5}{5!} = 0.1008.$$

***Example*** An insurance broker signs an insurance policy contract with the client on average three times a week. What is the probability that he will sell at least one insurance policy in a week?

Solution Since only one policy can be sold per unit of time and each unit of time is independent of the others, it is sufficient for the Poisson requirements that the broker's policy sales are proportional to the time.

To calculate the probability, it is enough to perform the following operations:

$$P(X > 0) = 1 - P(X = 0) = 1 - \frac{e^{-3}3^0}{0!} = 1 - 0.0498 = 0.9502.$$

*Note*: The average number of insurances sold per week is the same as the Poisson distribution parameter. That is $\lambda = 3$.

***Example*** Suppose 2% of the population has blood type *AB*. A sample of 60 people was randomly selected. If we consider the random variable $X$ as the number of people with blood type *AB*, what is the probability that none of them have blood type *AB*?

Solution It is clear that here $X$ is a random variable with a two-sentence distribution with parameters 60 and 0.02, meaning that $X \sim B(60, 0.02)$. To calculate this probability from a two-sentence distribution, we have to do calculations such as 60!, which will take a long time. But if we use the Poisson approximation, we can quickly come up with an almost perfect answer.

In this case, to approximate the two-sentence distribution, we use the Poisson distribution and say $X \sim P(60 \times 0.02 = 1.2)$. Now we get the probability $P(X = 0)$

$$P(X = 0) = \frac{e^{-\lambda}\lambda^x}{x!} = \frac{e^{-1.2} \times 1.2^0}{0!} = 0.3012.$$

But if we do these calculations based on the distribution of two sentences, we will get the following answer with a lot of time. This answer differs from the Poisson approximation by about 0.003:

$$P(X = 0) = \binom{60}{0} 0.02^0 \times 0.98^{60} = .2975.$$

## 2.7 Important Continuous Random Variables

In probability theory, distributions of random variables are of great importance for recognizing random phenomena. Among these, continuous uniform, gamma, exponential, and normal distributions have many applications, especially in simulating data related to other statistical distributions. Therefore, in this section, we will examine these statistical distributions and explain the characteristics of each.

### 2.7.1 *Continuous Uniform*

**Definition** Suppose $X$ is a continuous random variable whose domain is $(a, b)$. If the density function of this random variable is written as follows, we say that the random variable $X$ has a continuous uniform distribution, and we write $X \sim U(a, b)$.

$$f_X(x) = \begin{cases} \frac{1}{b-a}, & a \leq x \leq b \\ 0, & \text{o.w.} \end{cases}$$

The uniform probability distribution function and uniform cumulative probability distribution function are respectively represent in Figs. 2.17 and 2.18.

We see that $f_X(x) > 0$ and

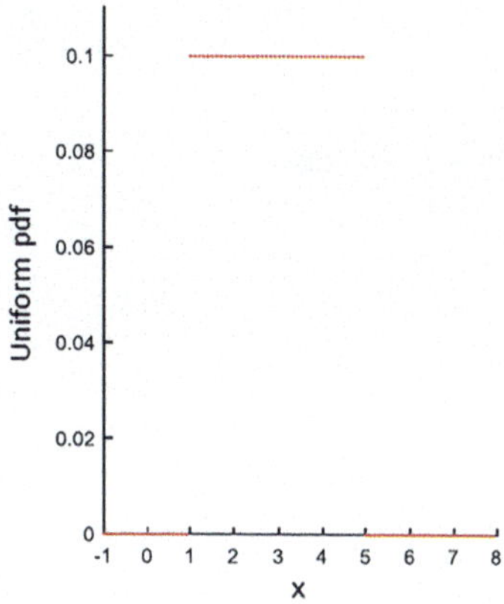


**Fig. 2.17** The uniform probability distribution function

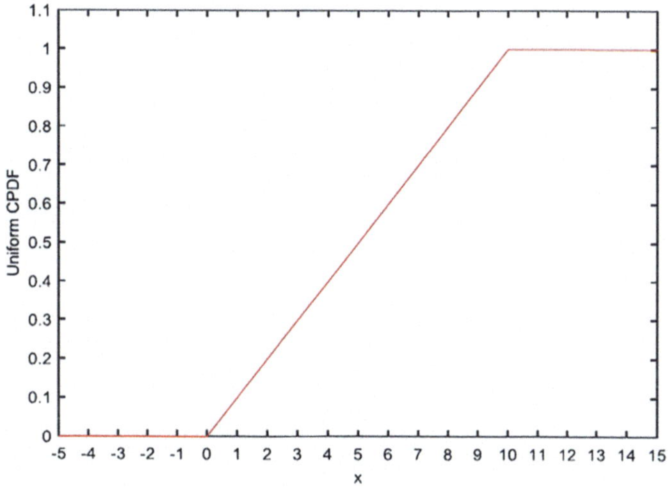


**Fig. 2.18** The uniform cumulative probability distribution function

$$\int_a^b \frac{1}{b-a} dx = \frac{b-a}{b-a} = 1.$$

Notice that from definition, the cdf of continuous uniform is expressed as follows:

$$F_X(x) = \begin{cases} \frac{x-a}{b-a}, & a \le x \le b \\ 0, & \text{o.w.} \end{cases},$$

**Theorem** *The expected value and variance of a uniform random variable X are given as follows:*

$$\mu_X := E(X) = \frac{b-a}{2} \text{ and } \sigma_X^2 := Var(X) = \frac{(b-a)^2}{12}.$$

***Proof*** We have

$$E(X) = \int_a^b x\left(\frac{1}{b-a}\right)dx = \left.\frac{x^2}{2(b-a)}\right|_a^b$$
$$= \frac{b^2-a^2}{2(b-a)} = \frac{b+a}{2}.$$

Now,

$$E(X^2) = \int_a^b x^2\left(\frac{1}{b-a}\right)dx = \left.\frac{x^2}{3(b-a)}\right|_a^b$$
$$= \frac{b^3-a^3}{3(b-a)} = \frac{(b-a)(b^2+ab+a^2)}{3(b-a)}$$
$$= \frac{b^2+ab+a^2}{3}.$$

Hence,

$$Var(X) = E(X^2) - E(X)^2$$
$$= \frac{b^2+ab+a^2}{3} - \left(\frac{b+a}{2}\right)^2 = \frac{(b-a)^2}{12}.$$

***Example*** Let $X$ be a uniform random variable on [1, 2]. Find $P(X > 3)$, $\mu_X$, $\sigma_X^2$, and standard deviation of $X$.

Solution According to the pdf of this uniform random variable, we have

$$f(x) = \begin{cases} \frac{1}{5}, & 1 \le x \le 6, \\ 0, & \text{o.w.} \end{cases}$$

Hence,

$$P(X > 3) = \int_3^\infty \frac{1}{5}dx = \int_3^6 \frac{1}{5}dx = \frac{3}{5} = 60\%.$$

The expected value of this random variable is

$$\mu_X = \frac{6+1}{2} = \frac{7}{2} = 3.5.$$

The second moment of $X$ is

$$\mu_X = \frac{b^2+ab+a^2}{3} = \frac{36+6+1}{3} = \frac{43}{3}.$$

Therefore,

$$\sigma_X^2 = \frac{43}{3} - \left(\frac{7}{2}\right)^2 = \frac{25}{12} \approx 2.08,$$

and

$$\sigma_X = \sqrt{2.09} = 1.44.$$

### 2.7.2 *Gamma*

In probability statistics and theory, gamma distribution is one of the functions of continuous distribution and it is known with three shape parameter $\alpha$, rate parameter $\lambda$ and location parameter $l$.

**Definition** Suppose that $\alpha$ and $\lambda$ are the positive real numbers. Define the pdf of random variable $X$ as follows:

$$f(x;\lambda;\alpha,l) = \begin{cases} \frac{\lambda^\alpha}{\Gamma(\alpha)}(x-l)^{\alpha-1}e^{-\lambda(x-1)}, & x \geq l, \alpha, \lambda > 0, \\ 0, & x < 0, \end{cases}$$

where $\Gamma(\alpha)$ is the gamma function given by

$$\Gamma(\alpha) = \int_0^\infty x^{\alpha-1}e^{-x}dx.$$

If $l = 0$, then the gamma random variable pdf with two parameters is obtained as follows:

$$f(x;\lambda;\alpha) = \begin{cases} \frac{\lambda^\alpha}{\Gamma(\alpha)}x^{\alpha-1}e^{-\lambda x}, & \alpha, \lambda > 0, x > 0, \\ 0, & x < 0, \end{cases}$$

where $\Gamma(\alpha)$ is the gamma function. Now, if we let $t = \lambda x$ and write $x = t/\lambda$, we will have:

$$\int_0^{\infty} \frac{\lambda}{\Gamma(\alpha)} t^{\alpha-1} e^{-t} dt = 1,$$

and the integrand is nonnegative. Thus, the three-parameter gamma random variable is a pdf.

If we let $\alpha = 1$, then the pdf of one-parameter gamma random variable is

$$f(x; \lambda; \alpha) = \begin{cases} \lambda e^{-\lambda x}, & \lambda > 0, x \geq 0, \\ 0, & x < 0. \end{cases}$$

**Theorem** *For a gamma random variable with two parameters X, the expected value and variance are given as follows:*

$$\mu_X := E(X) = \frac{\alpha}{\lambda} \text{ and } \sigma_X^2 := Var(X) = \frac{\alpha}{\lambda^2}.$$

***Proof*** We have:

$$E(X) = \int_{-\infty}^{\infty} x \frac{\lambda^{\alpha}}{\Gamma(\alpha)} x^{\alpha-1} e^{-\lambda x} dx = \int_0^{\infty} x \frac{\lambda^{\alpha}}{\Gamma(\alpha)} x^{\alpha-1} e^{-\lambda x} dx.$$

Substituting $t = \lambda x$ and write $x = \frac{t}{\lambda}$, hence, $dx = \frac{dt}{\lambda}$, we will have:

$$E(X) = \int_0^{\infty} \frac{t}{\lambda} \frac{\lambda^{\alpha}}{\Gamma(\alpha)} \left(\frac{t}{\lambda}\right)^{\alpha-1} e^{-t} \frac{dt}{\lambda} = \frac{1}{\lambda \Gamma(\alpha)} \int_0^{\infty} t^{\alpha} e^{-t} dt.$$

Using the property of the gamma function, the integral in the right-hand side is

$$\Gamma(\alpha + 1) = \alpha \Gamma(\alpha).$$

Hence, we will have:

$$E(X) = \frac{1}{\lambda \Gamma(\alpha)} \alpha \Gamma(\alpha) = \frac{\alpha}{\lambda}.$$

From the second moment definition, substituting $t = \lambda x$, and $x = \frac{t}{\lambda}$, we obtain

$$E(X^2) = \int_0^{\infty} x^2 \frac{\lambda}{\Gamma(\alpha)} (\lambda x)^{\alpha-1} e^{-\lambda x} dx = \int \frac{t^2}{\lambda^2} \frac{\lambda}{\Gamma(\alpha)} t^{\alpha-1} e^{-t} \frac{dt}{\lambda}$$

$$= \frac{1}{\lambda^2\Gamma(\alpha)} \int_0^\infty t^{\alpha+1} e^{-t} dt = \frac{1}{\lambda^2\Gamma(\alpha)} \Gamma(\alpha+2) = \frac{\alpha(\alpha+1)}{\lambda^2}.$$

Hence,

$$\sigma_X^2 = Var(X) = E(X^2) - E(X)^2 = \frac{\alpha(\alpha+1)}{\lambda^2} - \left(\frac{\alpha}{\lambda}\right)^2 = \frac{\alpha}{\lambda^2}.$$

### 2.7.3 Exponential

In a Poisson randomized trial, the number of events (success or failure) per unit time or place was considered. But if we consider the random variable as the time of the first event (success or failure), a continuous random variable is created that has an exponential distribution. For example, if the random variable is related to the waiting time of a customer to do banking or examining the length of the fabric to achieve the first flaw, it can be considered a random variable with exponential distribution. Also, most electrical and electronic components have a lifespan (time taken until burn the component) with an exponential distribution.

**Definition** If $X$ is a continuous random variable with domain of non-negative real number and its density function is written as follows, it has an exponential distribution:

$$f(x) = \lambda e^{-\lambda x}, x > 0.$$

In this case we write $X \sim E(\lambda)$ and read $X$ has an exponential distribution with parameter $\lambda$.

Of course, sometimes such a random variable is used to represent the density function (probability mass) of the following representation

$$f(x) = \frac{1}{\theta} e^{-\frac{1}{\theta}x}, x > 0.$$

It is clear that in the second form of the density function, there is the following relation between $\theta$ and $\lambda$:

$$\lambda = \frac{1}{\theta}.$$

In the Poisson distribution, we know that the average number of events (success or failure) per unit time is the same as the Poisson distribution parameter, $\lambda$. That is, in each unit of time, an average of $\lambda$ times of success is observed. As a result, one can expect an event to be observed over time $\lambda = \frac{1}{\theta}$.

Therefore, the average time to achieve the first success in the exponential distribution is equal to $\theta$. For this reason, the second form is sometimes used instead of the first form to represent the density function, so that the probability density function has a parameter that indicates the average length of time it takes to achieve the first success.

To find the exponential random variable distribution function, it is sufficient to take the integral probability density function. This calculates the area under the density function curve. So if we denote the probability distribution function by $F_X(y)$, we have:

$$\begin{aligned} F_X(y) &= \int_{-\infty}^{y} \lambda e^{-\lambda x} \\ &= 1 - e^{-\lambda y}. \end{aligned}$$

To obtain $P(X > y)$, it is sufficient to calculate the value of $e^{-\lambda y}$.

***Example*** Assuming it is $X \sim E(2)$, what is the probability that the random variable $X$ is greater than 4?

Solution

$$\begin{aligned} P(X > 4) &= 1 - P(X \leq 4) \\ &= 1 - (1 - e^{-2\times 4}) = 1 - 0.9997 = 0.0003. \end{aligned}$$

This example can be interpreted as follows. In a Poisson experiment that takes 2 units of time to achieve the first success, the probability that we will see the first success after 4 units of time will be very small (0.0003).

If $X \sim E(\lambda)$, then an integral can be used to obtain the expectation and variance of the exponential random variable.

$$E(X) = \frac{1}{\lambda} = \theta, \; Var(X) = \frac{1}{\lambda^2} = \theta^2.$$

As expected, the expectation, or the mean, of the random variable is equal to the parameter $\theta$. Also, the standard deviation (root of variance) for the exponential random variable will be equal to the value of expectation.

***Example*** On average, there is 1 crack in every 2 m of a 6-meter beam. If $X$ indicates the length of the beam that has no cracks, what is the probability that the beam does not have cracks?

Solution We know the average number for each meter of beam There is $\frac{1}{2}$ crack, so it makes sense to write $X \sim E(0.5)$. So we are looking for 6 m or more of beam length to get to the first crack. To calculate this probability, we have

$$P(X \geq 6) = 1 - F_X(6) = e^{-0.5\times 6} = 0.0498.$$

In this way, we are almost certain that the beam has cracks, because in only about 5% of the cases does the beam not have cracks.

***Example*** Assume telephone calls in a call center reach the rate of 6 per hour. If the waiting time for calling is a random variable:

I. The probability that the next call occurs in 5 to 15 min.
II. The mean, variance and standard deviation.
III. Moment generating function.

Solution. $X$ has the pdf of exponential with parameter $\lambda = 6$.

(I) The probability that the next call occurs 5 to 15 min:

$$P\left(\frac{1}{12} \leq X \leq \frac{1}{4}\right) = \int_{1/12}^{1/4} 6e^{-6x}dx$$

$$= -e^{-6x}\Big|_{1/12}^{1/4} = -\left(e^{-\frac{3}{2}} - e^{-\frac{1}{2}}\right) \approx 38.34\%.$$

(II)

$$\mu_X = E(X) = \frac{1}{\lambda} = \frac{1}{6} \approx 0.17 \text{ calls per hour.}$$

$$\sigma_X^2 = Var(X) = \frac{1}{\lambda^2} = \frac{1}{36} \approx 0.0278.$$

$$\sigma_X = \frac{1}{\lambda} = \frac{1}{6} \approx 0.17.$$

(III)

$$M_X(t) = \frac{6}{6-t},\ t < 6.$$

According to Bernoulli's random experiment, we know that there is memoryless property for a discrete random variable with a geometric distribution. This feature can also be checked for a continuous variable with exponential distribution. Thus, if we consider $t$ and $s$ as two times, we can write using conditional probability:

$$P(X \geq t + s|X > s) = P(X > t).$$

This relationship, based on the definition of conditional probability, shows that if we know that the waiting time for the first success is greater than $s$, the probability that this time is greater than $t + s$ depends only on $t$. In this case, if we consider $X$ as the life of a lamp (success means the lamp burns out), we can say that knowing that this lamp is 2,000 h old has no effect on its future life.

The above relationship can also be represented more simply:

$$P(X \geq t + s) = P(X > t)P(X > s).$$

Given this form of memoryless property, we can say that the probability that a lamp with an average lifespan of 100,000 h will work for more than 5000 h is equal to working for more than 2000 h first and then more than 3000 h. That is:

$$P(X \geq 5000) = e^{\left(-\frac{1}{10000}\times 5000\right)} = e^{-0.5} = 0.65065.$$

and also

$$\begin{aligned} P(X \geq 2000)P(X \geq 3000) &= e^{\left(-\frac{1}{10000}\times 2000\right)} e^{\left(-\frac{1}{10000}\times 3000\right)} \\ &= e^{-0.2}e^{-0.3} = e^{-0.5} = 0.6065. \end{aligned}$$

Due to the fact that electronic components have a lifespan with exponential distribution, the exponential random variable is used to determine the warranty period of such components. Consider the following example.

***Example*** Suppose an elevator company uses a command control circuit for its equipment with an average lifespan of 10 years. The company must offer a guarantee for the elevator. If the company guarantees the elevator control circuit for 4 years, what is the probability that it will have to replace the control circuit for free based on the warranty?

Solution Given the above explanation, it is clear that $X \sim E\left(\frac{1}{10}\right)$. Now we have to calculate the probability that the control circuit has a life of less than 4 years.

$$P(X < 4) = 1 - e^{-\frac{1}{10}\times 4} = 1 - 0.6703 = 0.3297.$$

Therefore, if the company chooses a 4-year warranty on the control circuit, it will have to replace the part free of charge approximately 33% of the time before the end of the warranty period, which may not be cost-effective. Now, in your opinion, how long is the warranty period that the company has to replace the part in maximum 10% of cases?

To answer this question, it is sufficient to perform the following operation based on the method of calculating the probability distribution function of the exponential random variable:

$$\begin{aligned} &\{(X \leq y) \leq 0.1 \rightarrow 1 - e^{-\left(\frac{y}{10}\right)} \leq 0.1 \\ &\rightarrow e^{-\left(\frac{y}{10}\right)} \geq 0.9 \rightarrow y \leq -10\ln(0.9) \rightarrow y \leq 1.054. \end{aligned}$$

As a result, if the warranty period is one year, it can be hoped that up to only 10% of the elevator control circuits will need to be replaced free of charge.

### 2.7.4 Normal

As mentioned, data distribution tells us how dispersed and extensive the data we collect is. Normal distribution is one of these patterns. For example, if you exercise every morning of the week, the distribution of your exercise time is linear. The data may be distributed in different ways depending on the nature of the experiment and various factors. Researchers often try to figure out which mathematical functions the data distribution is closer to, so that they can have a correct analysis of the nature of the distribution and the calculations on it. One of the most important statistical distributions is called the normal distribution. Another name for the normal distribution is natural distribution or the "Gaussian function" because this function was first proposed by Karl Friedrich Gauss. This distribution is one of the most important possible continuous distributions in probability theory. The reason for the nomenclature, as well as the importance of this distribution, is that many of the values obtained are consistent with the values obtained from this distribution during natural and physical fluctuations around a constant value. There are many things that follow a normal distribution, including the heights of people, the sizes of objects produced by industrial machinery, measurement errors, blood pressure readings, and test scores.

**Definition** Suppose $\mu$ is real nymber real number and $\sigma$ is a positive real number. Assume that $X$ is a continuous random variable and

$$\phi(x) = \frac{1}{\sigma\sqrt{2\pi}} e^{-\frac{(x-\mu)^2}{2\sigma^2}}, \ x, \mu \in (-\infty, \infty), \sigma > 0,$$

which is called the normal pdf. In this case, base on the following theorem $\phi(x)$ is a pdf and $X$ is called normal random variable.

**Theorem** *$\phi(x)$ is a pdf.*

***Proof*** Since $\sigma > 0$, $\phi(x)$ is positive. On the other hand,

$$I \equiv \int_{-\infty}^{\infty} \phi(x)dx = \frac{1}{\sigma\sqrt{2\pi}} \int_{-\infty}^{\infty} e^{-\frac{(x-\mu)^2}{2\sigma^2}} dx$$

Now, let $z = \frac{x-\mu}{\sigma}$. Then, $dz = \frac{dx}{\sigma}$ and.

$$I = \frac{1}{\sqrt{2\pi}} \int_{-\infty}^{\infty} e^{-\frac{z^2}{2}} dz.$$

Hence,

$$I^2 = \left( \frac{1}{\sqrt{2\pi}} \int_{-\infty}^{\infty} e^{-\frac{z^2}{2}} dz \right)^2$$
$$= \left( \frac{1}{\sqrt{2\pi}} \int_{-\infty}^{\infty} e^{-\frac{z^2}{2}} dz \right) \left( \frac{1}{\sqrt{2\pi}} \int_{-\infty}^{\infty} e^{-\frac{z^2}{2}} dz \right) = \frac{1}{2\pi} \int_{-\infty}^{\infty} e^{-\frac{(z+y)^2}{2}} dzdy.$$

Applying the polar coordinates: $z = r\cos\theta, y = r\sin\theta, r^2 = z^2 + y^2, \theta = \arctan\frac{y}{z}, dzdy = rdrd\theta, 0 < r < \infty$ and $0 \leq \theta \leq 2\pi$, we will have:

$$I^2 = \frac{1}{2\pi} \int_0^{2\pi} \int_0^{\infty} e^{-\frac{r^2}{2}} rdrd\theta = \frac{1}{2\pi} \int_0^{2\pi} \left( -\frac{e^{-r^2}}{2} \right)_0^{\infty} d\theta$$
$$= \frac{1}{2\pi} \int_0^{2\pi} d\theta = \frac{2\pi}{2\pi} = 1.$$

Thus, $I = 1$ and $\phi(x)$ is a pdf.

Figure 2.19 shows the normal pdf with various values of $\sigma$ and $\mu$.

**Definition** In mathematics, the error function is a function used in the sciences of probabilities, materials, statistics, and partial differential equations. The definition of this function is as follows:

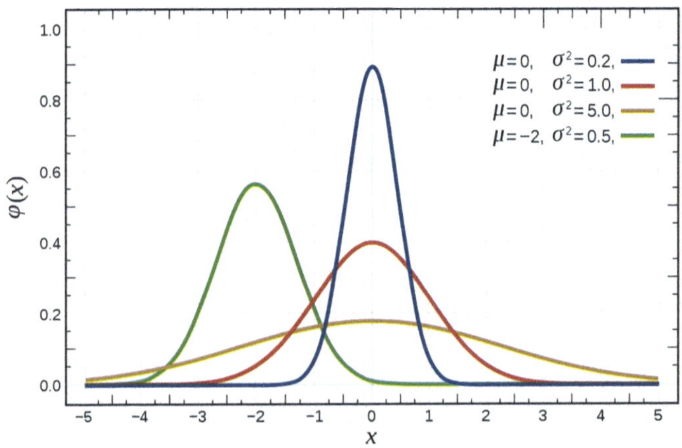


**Fig. 2.19** The normal pdf with various values of $\sigma$ and $\mu$

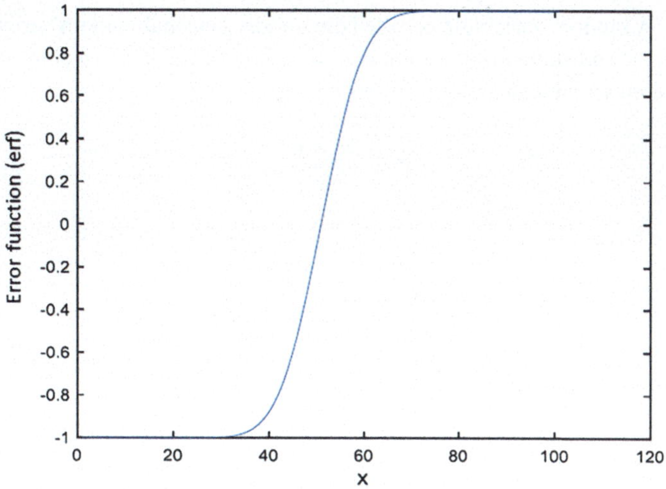


**Fig. 2.20** The diagram of erf for different values of $x$

$$erf\,(x) = \frac{1}{\sqrt{\pi}} \int_{-x}^{x} e^{-t^2} dt = \frac{2}{\sqrt{\pi}} \int_{0}^{x} e^{-t^2} dt.$$

Figure 2.20 indicates the diagram of error function (erf) for different values of $x$.

**Theorem** *For a normal or Gaussian random variable, the cdf, is denoted by* $\Phi(x; \mu, \sigma^2)$. *In this case, we have*

$$\begin{aligned} P(X \le x) = \Phi(x; \mu, \sigma^2) &= \int_{-\infty}^{x} \phi(t) dt \\ &= \int_{-\infty}^{x} \frac{1}{\sigma\sqrt{2\pi}} e^{-\frac{(t-\mu)^2}{2\sigma^2}} dt \\ &= \frac{1}{2}\left[1 + erf\left(\frac{x-\mu}{\sigma\sqrt{2}}\right)\right], \end{aligned}$$

Moreover,

$$E(X) = \mu,\ Var(X) = \sigma^2, \text{ and } M\,(t) = e^{\mu t + \frac{\sigma^2 t^2}{2}}.$$

***Proof*** Proof is given to the reader as an exercise.

The normal cdf is represented by $N(\mu, \sigma^2)$. In other words, the random variable $X$ with normal cdf is showed by $X \sim N(\mu, \sigma^2)$.

**Definition** A random variable $Z$ is called the standard normal random variable if the mean is $\mu = 0$ and standard deviation is $\sigma = 1$. The value of $Z$ is a conversion of a typical random variable defined as follows:

$$Z = \frac{X - \mu}{\sigma}.$$

Thus, we have

$$\phi(z) = \frac{1}{\sqrt{2\pi}} e^{-\frac{z^2}{2}}, \ -\infty < z < \infty,$$

Therefore, we can write

$$\Phi(z) = P(Z \leq z) = \int_{-\infty}^{z} \frac{1}{\sqrt{2\pi}} e^{-\frac{t^2}{2}} dt$$
$$= \frac{1}{2}\left[1 + erf\left(\frac{z}{\sqrt{2}}\right)\right].$$

***Example*** Suppose in a family the average height of people is 1.7 m and the standard deviation is 0.3 m. Assuming that the height of individuals follows a normal distribution, what is the value of z for a person whose height is 1.65 m?

Solution Due to this fact $\mu = 1.7$ and $\sigma = 0.3$, we have

$$z = \frac{x - \mu}{\sigma} = \frac{1.65 - 1.7}{0.3} = -0.166.$$

***Example*** When sugar packets enter the market from a factory, their average weight is 100 g with a variance of 25 g. Someone decides to buy a packet of sugar from a store. Calculate the following probabilities:

I. That the sugar package should be between 95 and 105 g.
II. The probability the sugar package is more than 105 g.
III. The probability of the sugar package is less than 90 g.

Solution

(I) Using the cdf of normal random variable, we have:

$$P(95 \leq X \leq 105) = \int_{95}^{100} \frac{1}{5\sqrt{2\pi}} e^{-\frac{(t-100)^2}{2.25}} dt \approx 0.6827.$$

(II) Again, using the cdf of normal random variable, we have:

$$P(X > 105) = \int_{105}^{\infty} \frac{1}{5\sqrt{2\pi}} e^{-\frac{(t-100)^2}{2.25}} dt = 1 - \int_{0}^{105} \frac{1}{5\sqrt{2\pi}} e^{-\frac{(t-100)^2}{2.25}} dt \approx 0.1587.$$

(III) Once more, using the cdf of normal random variable, we have:

$$P(0 \le X \le 90) = \int_{0}^{90} \frac{1}{5\sqrt{2\pi}} e^{-\frac{(t-100)^2}{2.25}} dt \approx 0.0228.$$

***Example*** Suppose the cost of repairing a car has a normal distribution with parameters $\mu = \$60$ and $\sigma = \$3$. What is the probability that the actual cost will be more than \$65?

Solution Assume X represent the cost of the car repair. In this case, we have

$$\begin{aligned} P(X > 65) &= P\left(X > \frac{65 - 60}{3}\right) = P\left(X > \frac{5}{3}\right) \\ &= 1 - \frac{1}{2}\left[1 + erf\left(\frac{5}{3\sqrt{2}}\right)\right] \approx 0.1587. \end{aligned}$$

***Example*** Suppose that $Z$ is a standard normal random variable. Find the following statements:

I. $P(-0.25 < z \le 0.46)$.
II. $P(0.87 < z \le 7.3)$.
III. $P(z < 6)$.
IV. $P(z > -0.40)$.

Solution

I. $P(-0.25 < z < 0.46) = \Phi(0.46) - \Phi(-0.25) = 0.6772 - 0.4013 = 0.2759$.
II. $P(0.87 < z < 7.3) = \Phi(7.3) - \Phi(0.87) = 1.000 - 0.8078 = 0.1922$.
III. $P(z < 0.46) = \Phi(6) = 1.0000$.
IV. $P(z > -0.40) = 1 - \Phi(-0.40) = 0.6554$.

## 2.8 Gambler's Ruin Problem

The earliest reference to the problem of the gambler's ruin relates to a letter from Blaise Pascal to Pierre de Fermat in 1656. Pascal's version of the problem is summarized in a letter from Pierre de Carcavi to Huygens in 1656:

Suppose two people play with three dice. The first person gets one point if 11 is thrown, and the second person gets one point if 14 is thrown. But instead of accumulating points in the usual way, one point is added to a player's score if his opponent's score is zero; otherwise, one point is deducted from his opponent. The winner is the player who gets 12 points. What is each player's relative chance of winning?

Huygens states the problem in *Libellus de Ratiociniis in Ludo Aleae*, 1657, as follows:

Each player starts with 12 points, and a successful throw of three dice for one player (first player 11 and second player 14) adds one point to his points and deducts one point from his opponent; the loser of the game is the first to reach zero points. What is the probability of victory for each player?

This is the classic expression of the gambler's ruin problem: two players start playing with certain starting points, and the points are transferred from one to the other until one of them reaches zero and "loses" the game. However, the term "gambler's ruin" was coined several years later.

More generally, the gambler starts with \$$i$ where $0 < i < N$. While the game proceeds,$\{R_n : n \geq 0\}$ forms a simple random walk

$$R_n = \Delta_1 + ... + \Delta_n,\ R_0 = i,$$

where $\{\Delta_n\}$ forms an independent and identically distributed (i.i.d.) sequence of random variables distributed as $P(\Delta = 1) = p$, $P(\Delta = -1) = q = 1 - p$, and represents the earnings on the successive gambles. Since the game stops when either $R_n = 0$ or $R_n = 1$, let

$$\tau_i = \min\{n \geq 0 : R_n \in \{0, N\} | R_0 = i\},$$

denote the time at which the game stops when $R_0 = i$. If $R_{\tau_i} = N$, then the gambler wins, if

$R_{\tau_i} = 0$, then the gambler is ruined.

Let $P_i = P(R_{\tau_i} = N)$ denote the probability that the gambler wins when $R_0 = i$. Clearly $P_0 = 0$ and $P_N = 1$ by definition, and we next proceed to compute $P_i$, $1 \leq i \leq N - 1$.

The key idea is to condition on the outcome of the first gamble, $P_N = 1$ or $P_N = 0$, yielding

$$P_i = pP_{i+1} + qP_{i-1}.$$

The derivation of this recursion is as follows: If $\Delta_1 = 1$, then the gambler's total fortune increases to $R_1 = i + 1$ and so, by the Markov property, the gambler will now win with probability $P_{i+1}$. Similarly, if $\Delta_1 = -1$, then the gambler's fortune decreases to $R_1 = i - 1$ and so, by the Markov property, the gambler will now win with probability $P_{i-1}$. Since $\mathrm{p} + \mathrm{q} = 1$, we can write

$$pP_i + qP_i = pP_{i+1} + qP_{i-1}.$$

Thus,

$$P_{i+1} - P_i = \frac{q}{p}(P_i - P_{i-1}).$$

In particular, $P_2 - P_1 = (q/p)(P_1 - P_0) = (q/p)P_1$ (since $P_0 = 0$), so that

$$P_3 - P_2 = \left(\frac{q}{p}\right)(P_2 - P_1) = \left(\frac{q}{p}\right)^2 P_1,$$

and more generally

$$P_{i+1} - P_i = \left(\frac{q}{p}\right)^i P_1, \ 0 < i < N.$$

Therefore,

$$\begin{aligned} P_{i+1} - P_1 &= \sum_{k=1}^{i} (P_{k+1} - P_k) \\ &= \sum_{k=1}^{i} \left(\frac{q}{p}\right)^k P_1, \end{aligned}$$

yielding

$$\begin{aligned} P_{i+1} &= P_1 + P_1 \sum_{k=1}^{i} \left(\frac{q}{p}\right)^k = P_1 \sum_{k=0}^{i} \left(\frac{q}{p}\right)^k \\ &= \begin{cases} P_1 \frac{1-\left(\frac{q}{p}\right)^{i+1}}{1-\left(\frac{q}{p}\right)}, & \text{if } p \neq q, \\ P_1(i+1), & \text{if } p = q = 0.5. \end{cases} \end{aligned}$$

Here we are using the "geometric series" equation $\sum_{n=0}^{i} a^i = \frac{1-a^{i+1}}{1-a}$, for any number a and any integer $i \geq 1$.

Choosing $i = N - 1$ and using the fact that $P_N = 1$ yields

$$1 = P_N = \begin{cases} P_1 \frac{1-\left(\frac{q}{p}\right)^{N}}{1-\left(\frac{q}{p}\right)} & \text{if } p \neq q, \\ P_1 N, & \text{if } p = q = 0.5. \end{cases}$$

from which we conclude that

$$P_1 = \begin{cases} P_1 \frac{1-\left(\frac{q}{p}\right)}{1-\left(\frac{q}{p}\right)^N} & \text{if } p \neq q, \\ \frac{1}{N} & \text{if } p = q = 0.5. \end{cases}$$

After algebraic simplification, we have

$$P_i = \begin{cases} P_1 \frac{1-\left(\frac{q}{p}\right)^N}{1-\left(\frac{q}{p}\right)^N} & \text{if } p \neq q, \\ \frac{i}{N} & \text{if } p = q = 0.5. \end{cases}$$

## 2.8.1 *Becoming Infinitely Rich or Getting Ruined*

Suppose that the gambler starting with $X_0 = i$ wishes to continue gambling forever until (if at all) ruined, with the intention of earning as much money as possible. Since there is no winning value $N$, the gambler will only stop if ruined. What will happen?

If $p > 0.5$, then $\frac{q}{p} < 1$ and thus

$$\lim_{N \to \infty} P_i = 1 - \left(\frac{q}{p}\right)^i > 0,\ p > 0.5.$$

This equation says that if p > 0.5 (each gamble is in his favor), then there is a positive probability.

that the gambler will never get ruined but instead will become infinitely rich.

If $p \leq 0.5$, then $\frac{q}{p} \geq 1$ and thus

$$\lim_{N \to \infty} P_i = 0,\ p \leq 0.5.$$

This equation says that if $p \leq 0.5$ (each gamble is not in his favor), then with probability one the gambler will get ruined.

***Example***

I. John starts with \$2, and $p = 0.6$ : What is the probability that John obtains a fortune of $N = 4$ without going broke?

*Solution* $i = 2,\ N = 4$ and $q = 1 - p = 0.4,\ q/p = 2/3,$ so and we want

$$P_2 = \frac{1 - \left(\frac{2}{3}\right)^2}{1 - \left(\frac{2}{3}\right)^4} = 0.91.$$

II. What is the probability that John will become infinitely rich?

*Solution* $1 - (q/p)^i = 1 - (2/3)^2 = 5/9 = 0.56.$

III. If John instead started with $i = \$1$, what is the probability that he would go broke?

*Solution* The probability he becomes infinitely rich is $1 - (q/p)^i = 1 - (q/p) = 1/3$, so the probability of ruin is $2/3$.

### 2.8.2 Random Walk Hitting Probabilities

Let $a > 0$ and $b > 0$ be integers, and let $R_n$ denote a simple random walk with $R0 = 0$. Let

$$p(a) = P(R_n \text{ hits level a befor hitting level } - b).$$

By letting $a = N - i$ and $b = i$ (so that $N = a + b$), we can imagine a gambler who starts with $i = b$ and wishes to reach $N = a + b$ before going broke. So we can compute $p(a)$ by casting the problem into the framework of the gamblers ruin problem: $p(a) = P_i$ where $N = a + b,\ i = b$. Thus,

$$p(a) = \begin{cases} \frac{1-\left(\frac{q}{p}\right)^b}{1-\left(\frac{q}{p}\right)^{a+b}}, & \text{if } p \neq q, \\ \frac{b}{a+b}, & \text{if } p = q = 0.5. \end{cases}$$

***Example***

I. Ellen bought a share of stock for \$10, and it is believed that the stock price moves (day by day) as a simple random walk with $p = 0.55$. What is the probability that Ellen's stock reaches the high value of \$15 before the low value of \$5?

*Solution* We want "the probability that the stock goes up by 5 before going down by 5." This is equivalent to starting the random walk at 0 with $a = 5$ and $b = 5$, and computing $p(a)$.

$$p(a) = \frac{1-\left(\frac{q}{p}\right)^b}{1-\left(\frac{q}{p}\right)^{a+b}} = \frac{1-(0.82)^5}{1-(0.82)^{10}} = 0.73.$$

II. What is the probability that Ellen will become infinitely rich?

*Solution* Here we keep $i = 10$ in the Gambler's ruin problem and let $N \to \infty$ in the formula for $P_{10}$.We have

$$\lim_{N\to\infty} P_{10} = 1 - \left(\frac{q}{p}\right)^{10} = 1 - (0.82)^{10} = 0.86.$$

## 2.9 Buffon's Needle Problem

A French nobleman, Georges Louis Leclerc, Comte de Buffon, posed the following problem in 1777 [3]:

> Suppose that you drop a short needle on ruled paper — what is then the probability that the needle comes to lie in a position where it crosses one of the lines?

The probability depends on the distance $d$ between the lines of the ruled paper, and it depends on the length $\ell$ of the needle that we drop—or rather it depends only on the ratio $\ell/d$. A *short* needle for our purpose is one of length $\ell \leq d$. In other words, a short needle is one that cannot cross two lines at the same time (and will come to touch two lines only with probability zero). The answer to Buffon's problem may come as a surprise: It involves the number $\pi$.

**Theorem** (Buffon's needle problem)
*If a short needle, of length $\ell$, is dropped on paper that is ruled with equally spaced lines of distance $d \geq \ell$, then the probability that the needle comes to lie in a position where it crosses one of the lines is exactly*

$$p = \frac{2}{\pi}\frac{\ell}{d}.$$

The result means that from an experiment, one can get approximate values for $\pi$ : If you drop a needle $N$ times, and get a positive answer (an intersection) in $P$ cases, then $P/N$ should be approximately $(2/\pi)(\ell/d)$, that is, $\pi$ should be approximated by $(2\ell N)/(dP)$. The most extensive (and exhaustive) test was perhaps done by Mario Lazzarini in 1901, who allegedly even built a machine in order to drop a stick 3408 times (with $= \ell/d = 5/6$). He found that it came to cross a line 1808 times, which yields the approximation $\pi \approx 2.(5/6)(3408/1808) = 3.1415929...$, which is correct to six digits of $\pi$ and much too good to be true! (The values that Lazzarini chose lead directly to the well-known approximation. This explains the more than suspicious choices of 3408 and $\frac{5}{6}3408$ is a multiple of 355.

The needle problem can be solved by evaluating an integral. We will do that below, and by this method we will also solve the problem for a long needle. But the Book Proof, presented by Joseph-Émile Barbier in 1860, needs no integrals. It just drops a different needle.

If you drop any needle, short or long, then the expected number of crossings will be

$$E = p_1 + 2p_2 + 3p_3 + \cdots ,$$

where $p_1$ is the probability that the needle will come to lie with exactly *one* crossing, $p_2$ is the probability that we get exactly two crossings, and $p_3$ is the probability for three crossings. etc. The probability that we get at least one crossing, which Button's problem asks for, is thus

$$p = p_1 + p_2 + p_3 + \ldots$$

Let $E(\ell)$ for the expected number of crossings that will be produced by dropping a straight needle of length $e$. If this length is $\ell = x + y$, and we consider the "front part" of length $x$ and the "back part" of length $y$ of the needle separately, then we get

$$E(x + y) = E(x) + E(y),$$

since the crossings produced are always just those produced by the front part, plus those of the back part.

By induction on $n$ this "functional equation" implies that $E(nx) = nE(x)$ for all $n \in \mathbb{N}$, and then that

$$mE\left(\frac{n}{m}x\right) = E\left(m\frac{n}{m}x\right) = E(nx) = nE(x),$$

Thus, $E(x) = cx$ for all $x \geq 0$, where $c = E(1)$ is some constant.

But what is the constant?

For that we use needles of a different shape. Indeed, let's drop a "polygonal" needle of total length $\ell$, which consists of straight pieces. Then the number of crossings it produces is (with probability 1) the sum of the numbers of crossings produced by its straight pieces. Hence, the expected number of crossings is again

$$E = c\ell,$$

by linearity of expectation. (For that result, it is not even important whether the straight pieces are joined together in a rigid or in a flexible way!)

The key to Barbier's solution of Baton's needle problem is to consider a needle that is a perfect circle $C$ of diameter $d$. which has length $x = d\pi$. Such a needle, if dropped onto ruled paper, produces exactly two intersections, always!

The circle can he approximated by polygons. Just imagine that together with the circular needle $C$ we are dropping an inscribed polygon $P_n$, *as* well as a circumscribed polygon $P^n$. Every line that intersects $P_n$, will also intersect $C$, and if a line intersects $C$ then it also hits $P^n$. Thus the expected numbers of intersections satisfy

$$E(P_n) \leq E(C) \leq E(P^n).$$

Now both $P_n$ and $P^n$ are polygons, so the number of crossings that we may expect is "$c$ times length" for both of them. while for $C$ it is 2, whence

$$c\ell(P_n) \leq 2 \leq c\ell(P^n).$$

Both $P_n$ and $P^n$ approximate $C$ for $n \to \infty$. In particular,

$$\lim_{n\to\infty} \ell(P_n) = d\pi = \lim_{n\to\infty} \ell(P^n),$$

and thus for $n \to \infty$. we have

$$cd\pi \leq 2 \leq cd\pi,$$

which gives $c = \frac{2}{\pi d}$.

But we could also have done it by calculus! The trick to obtain an "eas" integral is to first consider the slope of the needle. Let's say it drops to lie with an angle of $\alpha$ away from horizontal, where $\alpha$ will be in the range $0 \leq \alpha \leq \pi/3$. (We will ignore the case where the needle comes to lie with negative slope, since that case is symmetric to the case of positive slope, and produces the same probability). A needle that lies with angle $\alpha$ has height $\ell \sin\alpha$, and the probability that such a needle crosses one of the horizontal lines of distance $d$ is $(\ell \sin\alpha)/d$. Thus, we get the probability by averaging over the possible angles $\alpha$ as

$$p = \frac{2}{\pi} \int_0^{\frac{\pi}{2}} \frac{\ell \sin\alpha}{d} d\alpha = \frac{2}{\pi} \frac{\ell}{d} (-\cos\alpha)_0^{\pi/2} = \frac{2}{\pi} \frac{\ell}{d}.$$

For a long needle, we get the same probability $(\ell \sin\alpha)/d$ as long as $\ell \sin\alpha \leq d$, that is, in the range $0 \leq \alpha \leq \arcsin(d/\ell)$. However, for larger angles $\alpha$ the needle *must* cross a line, so the probability is 1. Hence, we compute

$$\begin{aligned} p &= \frac{2}{\pi} \int_0^{\arcsin(d/\ell)} \frac{\ell \sin\alpha}{d} d\alpha + \int_{\arcsin(d/\ell)}^{\pi/2} 1 d\alpha \\ &= \frac{2}{\pi} \left( \frac{\ell}{d} [-\cos\alpha]_0^{\arcsin(d/\ell)} \right) + \left( \frac{\pi}{2} - \arcsin(\frac{d}{\ell}) \right) \\ &= 1 + \frac{2}{\pi} \left( \frac{\ell}{d} \left( 1 - \sqrt{1 - \frac{d^2}{\ell^2}} \right) - \arcsin(\frac{d}{\ell}) \right), \end{aligned}$$

for $\ell \geq d$.

So the answer isn't that pretty for a longer needle, but it provides us with a nice exercise. Show ("just for safety") that the formula yields $2/\pi$ for $\ell = d$, that it is strictly increasing in $\ell$, and that it tends to 1 for $\ell \to \infty$.

**Table 2.8** The joint pmf for a spesified two random variables $X$ and $Y$

| | $Y = 2$ | $Y = 4$ | $Y = 5$ |
|---|---|---|---|
| $X = 1$ | $\frac{1}{12}$ | $\frac{1}{24}$ | $\frac{1}{24}$ |
| $X = 2$ | $\frac{1}{6}$ | $\frac{1}{12}$ | $\frac{1}{8}$ |
| $X = 3$ | $\frac{1}{4}$ | $\frac{1}{8}$ | $\frac{1}{12}$ |

## 2.10 Exercises

1. What is the coefficient of $y^{10}x^{60}$ in the expansion of $(2x^2 + y)^{40}$.
2. How many different ways can the letters of "UNIVERSITIES" word be arranged?
3. Find $P(X = 3)$ if $X$ is a continuous random variable.
4. If you ask one stranger on campus what day of the week he was born, what is the probability that he was born on Sunday?
5. In a classroom of 11 students, what is the probability of having at least two students that have the same

   (a) Day of birth?
   (b) Month of birth?

6. If $P(A) = 0.2,\ P(B) = 0.3$, find the possible value of $P(A\ and\ B)$ from below:

   (a) 0.25
   (b) 0.3
   (c) 0.1

7. Consider two random variables $X$ and $Y$ with joint pmf given in Table 2.8.

   (a) Find $P(X < 2,\ Y < 4)$.
   (b) Find the marginal PMFs of $X$ and $Y$.
   (c) Find $P(Y = 2|X = 1)$.
   (d) Are $X$ and $Y$ independent?

8. Let $X$ and $Y$ be as defined in Exercise 6. Consider a new random variable $Z = X - 2Y$.

   (a) Find the PMF of $Z$.
   (b) Find $P(X = 2|Z = 0)$.

(9) In term of $F_{X,Y}(x,\ y)$ calculate the following.

   (a) $P(X > n,\ Y < m)$.
   (b) $P(Y < m)$.

10. Consider the following joint pdf's for the random variable $X$ and $Y$. Circle the ones where $X$ and $Y$ are independent and cross out the other ones

$$f(x, y) = 4x^2y^3$$
$$f(x, y) = \frac{1}{2}(x^3y + y^3x)$$
$$f(x, y) = 6e^{-3x-2y}$$

11. Let $X$ and $Y$ each have range $\{1, 2, 3, 4\}$. The following formula gives their joint pmf

$$P(X = i, Y = j) = \frac{i + j}{80}.$$

Compute each of the following.

12. Let $X$ and $Y$ be two continuous random variables with a joint pdf

$$f(x, y) = \begin{cases} cx^2y(1 + y)\, 0 \leq x \leq 3 \text{ and } 0 \leq y \leq 3, \\ 0 \text{ otherwise.} \end{cases}$$

Find the value of $c$.

13. Let $X$ and $Y$ be two random variables random variables, and let $r$, $s$, $t$ and $u$ be real numbers.

    (a) Show that $Cov(X + 2,\ Y - 2) = Cov(X, Y)$.
    (b) Show that $Cov(2X,\ 3Y) = 6Cov(X,\ Y)$.
    (c) Show that $Cov(2X + 4,\ 3Y + 5) = 6Cov(X, Y)$.

14. If we assume $X$ is a continuous random variable with pdf $f(x)$ we consider the following transformation of $X$. They are

    (a) $Y = 2X + 3$(Linear).
    (b) $Y = |X|$(Absolute Value)
    (c) $Y = X^2$(Quadratic)
    (d) $Y = g(x)$(Monotonic)
    In each case, we will want to find $f_Y(y)$.

15. Let $X$ and $Y$ be two independent discrete random variables with the same CDFs $F_X(x)$ and $F_Y(y)$. Define

$$Z = \max(X, Y),$$
$$W = \min(X, Y).$$

Find CDFs of $Z$ and $W$.

16. Consider two random variables $X$ and $Y$ with joint PMF given by

$$P_{XY}(k, l) = \frac{1}{2^{k+l}}, \text{ for } k, l = 1, 2, 3, \ldots$$

17. Suppose that the number of customers visiting a fast food restaurant in a given day is $N \sim Poisson(\lambda)$. Assume that each customer purchases a drink with probability $p$, independently from other customers, and independently from the value of $N$. Let $X$ be the number of customers who purchase drinks. Let $Y$ be the number of customers that do not purchase drinks; so $X + Y = N$.

    (a) Find the marginal PMFs of $X$ and $Y$.
    (b) Find the joint PMF of $X$ and $Y$.
    (c) Are $X$ and $Y$ independent?

(d) Find $E[X^2Y^2]$.

18. If the Cumulative Distribution Function of the random variable $X$ is given by $F_X(x) = \frac{1+x}{2}$. $F_X(x) = \frac{1+x}{2}$

    (a) Find the maximum value of $x$.
    (b) Find the minimum value of $x$.
    (c) Find PDF of $x$.

19. Three couples attend a dinner. Each of the six people chooses randomly a seat at a row of chairs. What is the probability that no couple sits together?
20. Three couples attend a dinner. Each of the six people chooses randomly a seat at a round table. What is the probability that no couple sits together?
21. In a binary transmission channel, a bit 1 is transmitted with probability 0.6 and a bit 0 with probability $x$. The conditional probability of receiving a 1 given that a 1 was sent is 0.98, the conditional probability of receiving a 0 when a 0 was sent is 0.97.

    (a) What is the probability that a bit 1 was sent when receiving a 1?
    (b) What is the probability of error in transmission?
    (c) If every bit is transmitted independently of other bits, what is the probability of transmitting correctly of "0101"?

22. The number of cars arriving per minute at a toll booth on a particular bridge is Poisson distributed with a mean of 2.5. What is the probability that in any given minute,

    (a) No cars arrive?
    (a) Not more than two cars arrive?

23. For overseas flights, an airline has three different choices on its dessert menu—ice cream, apple pie, and chocolate cake. Based on past experience, the airline believes that each dessert is equally likely to be chosen.

    (a) If a random sample of four passengers is selected, what is the probability that at least two will choose ice cream for dessert?
    (b) If a random sample of 21 passengers is selected, what is the *approximate* probability that at least two will choose ice cream for dessert?

24. Let $X$ be a continuous random variable with the following PDF

$$f_X(x) = \begin{cases} ce^{-4x} & \text{if x} \geq 0, \\ b & otherwise. \end{cases}$$

    where c is a positive constant.
25. Let $X$ be a standard random variable, i.e. $X$ has PDF as follows:

$$f_X(x) = \frac{1}{\sqrt{2\pi}} e^{-\frac{x^2}{2}} \text{ for all } x \in \mathbb{R}.$$

Assume that $Y = \sqrt{|X|}$. Find $f_Y(y)$.

26. Let $X \sim Exponential(2)$ and $Y = 2 + 3X$.

    (a) Find $P(X > 2)$.
    (b) Find $E[Y]$ and $Var[Y]$.
    (c) Find $P(X > 2|Y < 11)$.

27. At a breakfast buffet, 93 people chose coffee and 47 people chose juice. 25 people chose both coffee and juice.

    If each person chose at least one of these beverages, how many people visited the buffet?

    One person is selected randomly from the visitors, what is the probability that he has chosen coffee and not juice?

28. The PDF of the random variable $X$ is given by

$$f_X(x) = \begin{cases} ax & \text{if x} \in [0, 2], \\ b & otherwise. \end{cases}$$

29. The error in the reaction time $X$ (in minutes) of a certain machine has the following distribution:

$$f_X(x) = \frac{2}{3}x^2, -1 \leq x \leq 1.$$

30. Consider the cumulative probability distribution of $X$ is as follows:

$$F(x) = \begin{cases} 0 & x < 1 \\ 0.2 & 1 \leq x < 4 \\ 0.45 & 4 \leq x < 6 \\ 0.6 & 6 \leq x < 10 \\ 1 & 10 \leq x \end{cases}$$

31. Suppose the number of years of experience ($X$) an employee possesses when joining a particular type of job is given by the following pmf:

$$P(X = x) = p(x) = \begin{cases} \frac{x}{10}, & x = 1, 2, 3, 4, \\ 0, & \text{else} \end{cases}$$

32. Suppose a random variable $X$ has a uniform distribution over $[1, \beta]$ for some $\beta > 1$. Calculate the value of $\beta$ if $P(2 < X < 5) = 0.5$.
33. Suppose a random variable $X$ has a uniform distribution over $[a, b]$, where $0 \leq a < b$. If the mean value of $X$ is 5 and variance is 3/4, what are the values of $a$ and $b$?
34. Let $X$ be a continuous random variable with pdf $f(x) = 1 - |x - 1|$, $0 \leq x \leq 2$.

    (a) Find cdf of $X$ and graph it.

(b) Find the first decile.
(c) Find 20th and 95th percentiles, and show them on the graph of cdf of $X$.
(d) Find the IQR.

## References

1. Scott M (2013) Applied stochastic processes in science and engineering. University of Waterloo, eBook
2. Tucker A (1994) Applied combinatorics. John Wiley & Sons Inc.
3. de Buffon GLL (1777) Essai d'arithmétique morale. Euvres philosophiques

# Chapter 3
# Stochastic Processes and Applications

One of the applications of probability, and, in fact, a generalization of random vector, which is vastly used in biology, cell biology, and birth-and-death (B-D) processes, is the concept of stochastic processes. Historically, after Andrey Kolmogorov started the modern probability theory, as we know it today, in 1933, it was interrupted by World War II. However, in the 1940s, stochastic or random processes, as applications of probability in potential theory, gained attention.

Before we move to a formal definition of stochastic process, we remind the readers from Chap. 2 that $(\Omega, S, P)$ is a probability space, where $P$ is a probability measure on a $\sigma$-field $S$.

**Definition** Let $T$ be a subset of non-negative real numbers, $[0, \infty)$. A stochastic (or random) process is a sequence (or a family) of random variables, say $X_t, t \in T$, indexed by $T$, defined on a probability space $(\Omega, S, P)$. The set $T$ is referred to as the index set or the parameter space. If $T$ is the set of non-negative integers, $\mathbb{N}_0$, the stochastic process is referred to as the discrete-time, denoted by $\{X(t), t \in T\}$ and if it is an interval $[0, \infty)$, it is referred to as the continuous-time, denoted by $\{X(t), t \in T\}$, stochastic process. If the probabilistic rules do not change with time, the process is called the stationary process.

**Definition** For a stochastic process $\{X_t, t \in T\}$ or $\{X(t), t \in T\}$, the set of all possible values of random variable $X_t$ or $X(t)$, that is, the value of a stochastic process at any epoch of time, denoted by $S$, is referred to as the state space of the process in discrete case and the phase space in continuous case, state space in both cases, if there is no confusion, which is a mathematical space like a subset of Euclidean $n$-dimensional space. This space reflects the different values the stochastic process can take. The state at which the process starts at, denoted by $X_0$ or $X(0)$, is referred to as the initial state.

It should be noted that if $T$ is finite such as $T = \{1, 2...., n\}$, we can find probability density function (pdf) or probability mass function (pmf) of the process through the



S. Brahim Belhaouari et al., *Discrete Mathematics, Probability Theory and Stochastic Processes*, Modeling and Optimization in Science and Technologies 20,
https://doi.org/10.1007/978-3-031-80589-9_3

joint distribution of random vector $\{X_{t_1},X_{t_2}, ..., X_{t_n}\}$. However, for an infinite case of $T$, finding pdf or pmf is somewhat problematic.

When the index set, say $T$, consists of only one element, that is, it is a singleton, $T = \{1\}$, the process $\{X_t, t \in T\}$ becomes a single random variable $X_1$. On the other hand, when $T$ is a finite set such as $T = \{1, 2, ...n\}$, the process will become a random vector, that is, $\{X_{t_1},X_{t_2}, ..., X_{t_n}\}$.

***Example*** Consider a random sample $\{X_1, X_2, ..., X_n\}$ of size $n$ from a population with an unknown distribution function. The function $F_n(x)$ referred to as the empirical distribution function is defined as

$$F_n(x) = \frac{Number\ of\ sample\ X_j \leq x}{n} = \frac{1}{n}\sum_{j=1}^{n} I_x(X_j), \ -\infty < x < \infty$$

where $I_x$ is the indicator function. Since $F_n(x)$ is a random variable for each $x$, the sequence $\{F_n(x), -\infty < x < \infty\}$ is a stochastic process.

***Example*** As an example of a discrete index set, consider the net number count of deliveries and returns of a particular item in an online store at time $t$ within 24 h of operation. Hence, the process of delivering and returning is a stochastic process with the index set $T$ as the set $\{0, 1, 2, ..., 24\}$ and with the state space as $S = \{0, \pm 1, \pm 2, ...\}$.

***Example*** As an example of a continuous index set, let us consider the arrivals of customers at a bank, which is assumed to be random in nature. The number of customers arrived during the interval $[0, t]$ is a stochastic process in which the index set is $T = [0, \infty)$ and the state space $S = \{0, 1, 2, ...\}$. This is also an example of a pure birth process, which we will consider later.

***Example*** Suppose $X_1, X_2, ...$ are independent random variables with $P(X_k = 1) = P(X_k = -1) = 1/2$ for each $k$, $S_0 = 0$, and $S_n = X_1 + ... + X_n$ for positive integers $n$. Let $X = (X_k : k \geq 1)$ and $S = (S_n : n \geq 0)$. Then $X$ and $S$ are both discrete-time random processes. The index set $T$ for $X$ is $\mathbb{Z}^+$. A sample path of $X$ and a corresponding sample path of $S$ are shown in Fig. 3.1.

The following notation is used:

| | |
|---|---|
| Mean function: | $\mu_X(t) = E[X_t]$, |
| correlation function: | $R_X(s, t) = E[X_s X_t]$, |
| Covariance function : | $C_X(s, t) = Cov(X_s, X_t)$ |
| nth order cumulative | $F_{X,n}(x_1, t_1; ...; x_n, t_n)$ |
| distribution function (CDF): | $= P(X_{t_1} \leq x_1, ..., X_{t_1} \leq x_n)$, |

***Example*** Suppose $A$ and $B$ are independent standard normal random variables. Define $X_t = A + Bt + t^2$ for all $t \in \mathbb{R}$. Let us describe the sample functions, the mean, correlation, and covariance functions, and the first and second order pdf's of $X$.

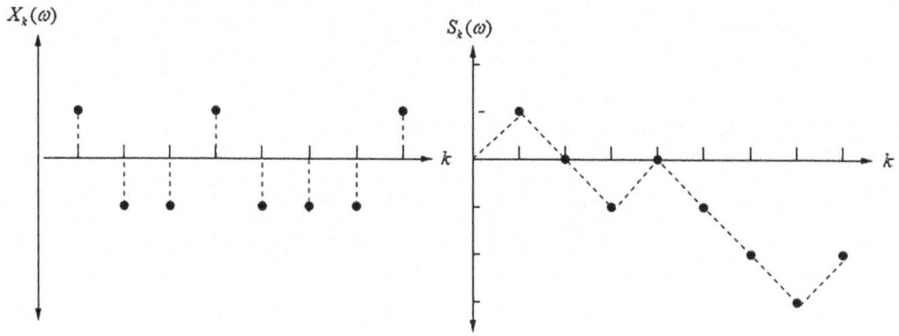


**Fig. 3.1** A sample path of discrete-time random process $X$ and a corresponding sample path of $S$

**Fig. 3.2** Three typical sample paths

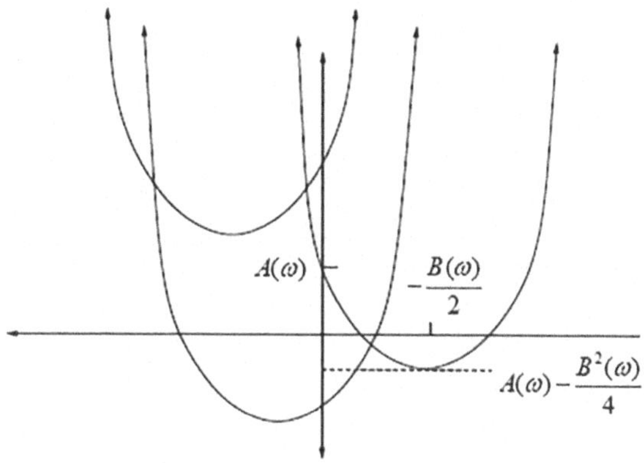


Each sample function corresponds to some fixed $\omega$ in $\Omega$. For $\omega$ fixed, $A(\omega)$ and $B(\omega)$ are numbers. The sample paths all have the same shape: they are parabolas with constant second derivative equal to 2. The sample path for $\omega$ fixed has $t = 0$ intercept $A(\omega)$, and minimum value $A(\omega) - \frac{B(\omega)^2}{4}$ achieved at $t = -\frac{B(\omega)}{2}$. Three typical sample paths are shown in Fig. 3.2.

The various moment functions are given by

$$\mu_X(t) = E[A + Bt + t^2] = t^2,$$
$$R_X(s,t) = E[(A + Bs + s^2)(A + Bt + t^2)] = 1 + st + s^2t^2,$$
$$C_X(s,t) = R_X(s,t) - \mu_X(s)\mu_X(t) = 1 + st.$$

As for the densities, for each $t$ fixed, $X_t$ is a linear combination of two independent Gaussian random variables, $\mu_X(t) = t^2$ and $Var(X_t) = C_X(t,t) = 1 + t^2$. Thus, $X_t$ is a $N(t^2, 1+t^2)$ random variable. That specifies the first order pdf $f_{X,1}$ well enough, but if one insists on writing it out in all detail it is given by

$$f_{X,1}(x,t) = \frac{1}{\sqrt{2\pi(1+t^2)}} \exp\left(-\frac{(x-t^2)^2}{2(1+t^2)}\right).$$

For distinct $s$ and $t$; $X_s$ and $X_t$ are jointly Gaussian with

$$Cov\begin{pmatrix} X_s \\ X_t \end{pmatrix} = \begin{pmatrix} 1+s^2 & 1+st \\ 1+st & 1+t^2 \end{pmatrix}.$$

The determinant of this matrix is $(s-t)^2$, which is nonzero. Thus $X$ has a second order pdf $f_{X,2}$ for most purposes, we have already written enough about $f_{X,2}$ for this example, but in full detail $f_{X,2}(x, s; y, t)$ is given by

$$\frac{1}{2\pi|s-t|} \exp\left(-\frac{1}{2}\begin{pmatrix} x-s^2 \\ y-t^2 \end{pmatrix}^T \begin{pmatrix} 1+s^2 & 1+st \\ 1+st & 1+t^2 \end{pmatrix}^{-1} \begin{pmatrix} x-s^2 \\ y-t^2 \end{pmatrix}\right).$$

The $n$th order distributions of $X$ are joint Gaussian distributions, but densities don't exist for $n \geq 3$ because the values of $\begin{pmatrix} X_{t_1} \\ X_{t_2} \\ X_{t_3} \end{pmatrix}$ are restricted to a plane embedded in $\mathbb{R}^3$.

## 3.1 Gaussian Process

The Gaussian Process in probability theory is a stochastic process known as a sequence of random variables. In this process, random variables are usually arranged in terms of time, and each finite subset of these variables has a Gaussian (normal) multivariate distribution.

So if $\{X_t; t \in T\}$ is a random sequence then $X_{t_1,t_2,\dots,t_k} = (X_{t_1}, \dots, X_{t_k})$ has a normal (Gaussian) multivariate distribution. In other words, in a Gaussian process, each linear combination of a finite set of processes has a normal univariate distribution. It is clear that $T$ is an infinite set of time or space indices. This process is named after the work and efforts of the German scientist Carl Friedrich Gauss in this field. To better describe the Gaussian model and process, we will give an example.

Suppose you are hired to coach a basketball team. Of course, you are also a data scientist. The basketball coach's motto can be summed up in three general guidelines: commit fewer turnovers, take more shots, and don't forget the team's cooperation in scoring points.

This statement can also be written based on statistical analysis and optimization as follows:

Dimension 1: The number of shots must be maximized.

Second dimension: The number of errors leading to the loss of the ball must be minimized.

Output or Objective Function: Maximize points per game.

It is clear that the first and second dimensions include the constraints by which the output or function of the target is obtained. We now examine and model the scores of a particular player (for example, Steph Curry) in terms of the number of shot attempts and turnovers per game. Thus, it seems that we are facing a two-dimensional Gaussian process and scores are considered as a dependent variable of this process.

The two-dimensional process can be described in terms of its variables. In the first dimension is the number of shots and in the second dimension is the loss of the ball. These variables will act as "regressor" variables in the created model. Of course, the output variable, which is the same as the score obtained in each game, will be used as a dependent variable in the model structure. First, we standardize the information obtained from these variables so that by calculating the zscore, the data will have a mean of zero and a variance of 1. For this reason, on the horizontal and vertical axis in the following diagrams, the scales change between − 3 to + 3 (due to small changes on the vertical axis, only the area between − 2 to + 2 is displayed.) This property in variable values the target or dependent can also be viewed.

In this way, the following images show a description of the relationship between independent and dependent variables based on the "contour plot". For example, a value of 1 in these charts indicates a difference of one standard deviation from the NBA average. The green dot inside the graphs also represents the actual value observed. Thus, the blue area indicates a better prediction for the score by the predictor variables. The further we go into the red zone, the more deviant the prediction becomes. On the other hand, the chart guide, which shows the color range from red to bright blue, shows his score in the game based on standardized values. In order to analyze the scores and performance of this player, first based on one game, then with the help of his last two games, and then based on the last three and four games, we infer and analyze. We expect that the more games there are, the more information and data will be about the player's performance.

Analysis of the first game: As can be seen, this player is very close to the average number of shots and turnovers in his first game. On the other hand, the points earned by him are a little more than the standard. It is clear that all estimates are based on only one point, so there does not seem to be a deviation.

Analysis of the third game: The number of shots by the basketball player has decreased and on the other hand, his errors in losing the ball have also increased. Therefore, it can be seen that the points obtained in this competition have decreased because the area related to this competition is shown in orange. As a result, he seems to have dropped compared to previous races. It seems that modeling in this case is suitable for predicting scores in the second game, but the prediction error for scoring in the third game is high.

Fourth game analysis: Due to the increase in the number of shots of the player, the points earned in the fourth game has not increased because the number of turnovers also shows an increase. Thus, predicting the score from game four is also a big mistake. Also, the second game and the first game are well estimated.

If we continue in the same way, we will see that there is a relationship between dependent and independent or descriptive variables. For example, when a player has a large number of turnovers, even if he has good number of shots, his score will be lower than the NBA average. But according to the graphs, it seems that our player is more efficient than the average (zero value for points) because his average score in these four matches is more than zero.

## 3.2 Ergodic Process

In probability theory, "random processes" are phenomena that change by chance over a long period of time. "Probability Distribution" is often used to express the chance or probability that such processes will shift or change. These stochastic processes are classified according to their characteristics. One of the interesting features of some processes is that they have an "ergodic" property, in which case it is called an ergodic process [1]. In this book, we want to know what the ergodic process is and what its characteristics are.

In the field of "econometrics" as well as "signal processing", if the statistical properties of the process can be determined on the basis of a random sample over a sufficiently long period of the process, it is called "Ergodic Process".

Each random sample of random processes may represent the average of the total characteristics of that process. In other words, this feature shows that regardless of the whole process, its behavior can be determined with the help of a sample. The presence of such a property in the random process determines "ergodicity."

By formal definition, it can be said that a random process is ergodic if each set of random samples of that process can show the average of the statistical characteristics of the whole process. In other words, no matter which random instances we use, by focusing on just one instance of the process, we are able to display the behavior of the whole process and its characteristics. Conversely, a non-ergodic process is a process whose changes have an uncertain and irregular rate, and the resulting samples cannot determine the overall behavior of the process. At the end of this text, we will give examples of ergodic and non-ergodic processes.

There are several ways to express the ergodic nature of a process, one of which is "wide-sense stationary" or long-term. For example, consider the random process $X(t)$, which has a fixed mean,

$$\mu_X = E[X(t)].$$

Also consider the "serial covariance" (autocovariance) as follows:

$$r_X(\tau) = E[(X(t) - \mu_X)(X(t+\tau) - \mu_X)]$$

The above relations show that the relationship between variables, over time with interval or time interval $\tau$ and it does not depend on time $t$.

Process $X(t)$ is called "Mean-Ergodic" or "Means-Square Ergodic in the First Moment" if at the time $T \rightarrow \infty$ is the average time estimate of the process that converges to the overall average by the following equation specified in the "Squared Mean":

$$\hat{\mu}_X = \frac{1}{T}\int_0^T X(t)dt.$$

In contrast, a process is called "autocovariance-ergodic" or in terms of torque $d$ if, when $T \rightarrow \infty$ goes, the average estimate converges to the overall average $r$. You can see the average time estimate in the following relation.

In contrast, a process is called "autocovariance-ergodic" or in terms of torque $d$ if, when $T \rightarrow \infty$ goes, the average estimate converges to the overall average $r_X(\tau)$. We can see the average time estimation is satisfied in the following relationship:

$$\hat{r}_X = \frac{1}{T}\int_0^T [(X(t+\tau) - \mu_X][X(t) - \mu_X]dt.$$

We have considered continuous values. Now we assume this parameter to be discrete (integer values). In this way, the term ergodic can also be used for "Discrete-time Random Process". Obviously in this case $X_{[n]}$ for integers $n$ is a time-discrete process. Note that the following equation holds for such a process

$$\hat{\mu}_X = \frac{1}{N}\sum_{n=1}^{N} X_{[n]}.$$

Remember that the above relation, when $N \rightarrow \infty$, converges to $E[X]$. $E[X]$.

As mentioned earlier, the meaning of "ergodicity" means that the overall average is equal to the average time. The following examples are provided to illustrate this principle.

**Telephone Center**

Each operator in the call center alternates between talking and listening to the phone, and there is also a break between calls. Each break and each call have a different length of time. Also, the duration of each communication (talking and listening) and the speed of communication at any given moment can be modeled as a random process.

- In this example, consider the number of call center operators $N$ ($N$ must be a very large integer) and plot the number of words spoken per minute for each operator over a long period (several work shifts). We will have a series of points for each operator that can be used to create a waveform plot.

- Calculate the average value of those points in the waveform. This gives you the average time.
- In this way, you will have $N$ wave diagrams for the number of $N$ operators. These wave diagrams are known as a whole or group.
- Now specify a specific moment or time in all those wave charts and find the average value of the number of words spoken at that time. This calculation gives you the group average for that moment.
- If the group average is always equal to the average time, the system has ergodic properties.

**Electronic Resistors**

Each resistor has a thermal noise that depends on the temperature. Test the resistance at $N$ ($N$ must be very large) and maintain a voltage between those resistors for a long time. Draw a diagram for the output voltage of these resistors. The output voltage creates a graph in the form of a wave. Calculate the average voltage for each of the resistors. This gives you the average time. Since $N$ is a resistor, we will know it as a group. Now consider a specific moment or time interval in all those diagrams and find the average value of the voltages on the diagrams. This gives you the group average for each chart. If the group average and the average time are the same, then the process of voltage changes is an ergodic process.

In the continuation of this section, we refer to random processes that are non-ergodic and do not have ergodic properties.

Random Walk, which is done randomly, is non-ergodic. Its expected value or mathematical expectation is zero at all times, while its mean time is a variable with divergent variance.

Suppose we have two coins. One coin is head/tail coin and the other coin is head on both sides. First we select one of the coins (randomly) and then we perform a sequence of independent throws of the selected coin. Suppose $X_{[n]}$ is the $n$th result of this coin toss, in which we have a value of 1 for the head and 0 for the tail. Now the average value observed (for time period $n$ is equal to $\frac{1}{2}\left(\frac{1}{2}+1\right)$, of which $\frac{3}{4}$ is obtained as a result. The first will have a probability of $\frac{1}{2}$ head, and the second coin will have a probability of 1 result, while if a non-odd coin is selected, the average number of heads in the long run will be $\frac{1}{2}$ and for the other coin will be 1. The average time is not equal to the group average, so such a process would not be an ergodic process.

**Definition** Let $X_1, X_2, \ldots,$ be a sequence of i.i.d random variables. Also, let $S_n = X_1 + X_2 + \ldots + X_n$. Then, the sequence of sums, $\{S_n, n \geq 1\}$, which is the integer-time stochastic process, is referred to as a simple random walk or one-dimensional random walk, based on $\{X_j, j \geq 1\}$. The word "simple" will be dropped when higher dimensions are considered.

***Remark*** The increments $X_k - X_{k-1}, k = 1, 2, \ldots, n$, are independent of $\{X_n\}, n = 1, 2, \ldots$. Also, each increment $X_k - X_{k-1}, k = 1, 2, \ldots, n$, has a Bernoulli pmf with the probability of success as $p$. That is,

$P(X_k - X_{k-1}) = p$ and $P(X_k - X_{k-1}) = 1 - p$, $k = 1, 2, ..., n$, $n = 1, 2, ..., X_0 = 0$.

**Definition** Let $\tau > 0$ be a positive real number. A term in $\{S_n, n \geq 1\}$ for which $S_n \geq \tau$ is referred to as the threshold at $\tau$.

***Example*** Imagine a walker who starts at point 0, walking randomly one step (with the same length throughout the process) forward (+1) and one step backward (−1). That is, staying put is not allowed. Suppose he stops after ten steps. During the process of walking, if he walked four steps forward, he will be at position $S_{10} = -2$.

Note that if we allow him to stay put during the walking process, we can assign 0 to values of the random variables.

***Example*** In this example, we will show how we may construct a simple random walk process. We start with the probability space $(\Omega, S, P)$ and the stochastic process $\{X_k, k = 1, 2, ...\}$. So, we suppose that $X_1, X_2, ...$ are i.i.d. random variables, each with two values + 1 and − 1, with probabilities $p$ and $1 - p$, respectively. Let $S_n = X_1 + X_2 + .... + X_n$. Now if there are $j$ positive steps out of the $n$ total steps, then $S_n = 2j - n$ (why?). Hence,

$$P(S_n = 2j - n) = \frac{n!}{j!(n-j)!} p^j (1-p)^{n-j}.$$

So, the probability of position of the walker having four forward steps, with the 2/3 probability of each such a step as, in a total of ten steps, is:

$$P(S_{10} = -2) = \frac{10!}{4!(6)!} \left(\frac{2}{3}\right)^j \left(\frac{1}{3}\right)^6 = 0.0569.$$

Hence, the position of the walker depends upon the value of $\overline{X}$ as it is negative, 0, or positive. If $\overline{X} > 0$, then $S_n$ will tend to drift downward, and if $\overline{X} < 0$, then $S_n$ will tend to drift upward. Of course, position for $\overline{X} > 0$ can be obtained from the $\overline{X} < 0$ by the use of $\{-S_n, n \geq 1\}$.

***Example*** Two gamblers, designated by $G_1$ and $G_2$ (Gambler 1 and Gambler 2, respectively), play a simple gambling game. On each round (trial or betting) of the game, a coin is tossed, and if $H =$ "heads" results, then $G_1$ wins \$1 from $G_2$. On the other hand, if $T =$ "tails" results from the coin tossing, then $G_2$ wins \$1 from $G_1$ The probabilities associated with "heads" and "tails" on the coin are unknown to the players so that both players feel free to continue the rounds. The game continues until one of the players loses all his/her money and is thereby "ruined".

Almost all the properties listed are properties of the one dimensional distributions of $X$. In fact, only the strong law of large numbers, giving the a.s. convergence in the third property listed, depends on the joint distribution of the $X_n$'s.

The gambler's ruin problem is a nice example of the calculation of a probability involving the joint distributions of the random walk $X$. Interpret $X_n$ as the number of units of money a gambler has at time $n$. Assume that the initial wealth $k$ satisfies

$k \geq 0$, and suppose the gambler has a goal of accumulating $b$ units of money for some positive integer $b \geq k$. While the random walk $(X_n : n \geq 0)$ continues on forever, we are only interested in it until it hits either 0 (the gambler is ruined) or $b$ (the gambler is successful). Let $S_b$ denote the event that the gambler is successful, meaning the random walk reaches $b$ or more without first reaching 0. The gambler's success probability is $P_k(S_b)$.

Let $s_k = P_k(S_b)$ for $0 \leq k \leq b$, so $s_k$ is the success probability for the gambler with initial wealth $k$ and target wealth $b$. Clearly $s_0 = 0$ and $s_b = 1$. For $1 \leq k \leq b-1$, condition on $W_1$ to yield

$$s_k = P_k(W_1 = 1)P_k(S_b|W_1 = 1) + P_k(W_1 = -1)P_k(S_b|W_1 = -1)$$

or $s_k = ps_{k+1} + (1-p)s_{k-1}$. This yields $b-1$ linear equations for the $b-1$ unknowns $s_1, \ldots, s_{b-1}$.

If $p = \frac{1}{2}$, the equations become $s_k = \frac{1}{2}(s_{k-1} + s_{k+1})$ so that $s_k = A + Bk$ for some constants $A$ and $B$. Using the boundary conditions $s_0 = 0$ and $s_b = 1$, we have $s_k = \frac{k}{b}$ in case $p = \frac{1}{2}$. Note that, interestingly enough, after the gambler stops playing, he'll have $b$ units with probability $\frac{k}{b}$ and zero units otherwise. Thus, his expected wealth after completing the game is equal to his initial capital, $k$.

If $p \neq \frac{1}{2}$, we seek a solution of the form $s_k = A\theta_1^k + B\theta_2^k$ where $\theta_1$ and $\theta_2$ are the two roots of the quadratic equation $\theta = p\theta^2 + (1-p))$ and A, B are selected to meet the two boundary conditions. The roots are 1 and $\frac{1-p}{p}$ and finding A and B yields that, if $p \neq \frac{1}{2}$,

$$s_k = \frac{1 - \left(\frac{1-p}{p}\right)^k}{1 - \left(\frac{1-p}{p}\right)^b}, \quad 0 \leq k \leq b.$$

Now suppose $p > \frac{1}{2}$. By the law of large numbers, $\frac{X_n}{n} \to 2p - 1$ a.s. as $n \to \infty$. This implies, in particular, that $X_n \to +\infty$ a.s. as $n \to \infty$. Thus, unless the gambler is ruined in finite time, his capital converges to infinity. Let $S$ be the event that the gambler's wealth converges to infinity without ever reaching zero. The events $S_b$ decrease with $b$, because if $b$ is larger, the gambler has more possibilities to be ruined before accumulating $b$ units of money: $S_1 \supset S_2 \supset \ldots$ and $S = \{X_n \to \infty\} \cap (\cap_{b=1}^{\infty} S_b)$. Therefore, by the fact $P(X_n \to \infty) = 1$ and the continuity of probability,

$$P_k(S) = P(\cap_{b=1}^{\infty} S_b) = \lim_{b \to \infty} P_k(S_b) = \lim_{b \to \infty} s_k = 1 - \left(\frac{1-p}{p}\right)^k.$$

Thus, the probability of eventual ruin decreases geometrically with the initial wealth $k$.

## 3.3 Point Process

As another case of a stochastic process, we briefly discuss the point process.

**Definition** A set of random points $\{T_1, T_2, ..., T_r\}$ such that $T_1 < T_2 < ... < T_r$, $\lim_{r\to\infty} T_r = \infty$, in some space, often a subset of multidimensional Euclidean space, say $\mathbb{R}^r$, for some $r$, is a stochastic process referred to as a point process. In $\mathbb{R}^1$, each $T_j$ is a random variable. The numbers $T_r$, $r = 1, 2, ...$, are referred to as the event times or epochs.

According to the definition above, a point process is a strictly increasing sequence of real numbers, which does not have a finite limit point.

As examples of point process, we can mention the arrival time epochs of customers at a service station and in biology by the moments of impulses in nerve fibers.

We note that $T_r$, $r = 1, 2, ...$, do not necessarily have to be points of time. They could be points other than times like location of potholes in a road. However, since the number of potholes is finite, in the real world, to satisfy the conditions stated in the definition of threshold, we need to consider finite sample from a point process. The arrival times are of less interest than the number arrived in an interval of time.

**Definition** Let $T_r, r = 1, 2, ...$, represent the number of events occurred in an interval of time $(0, T]$, $t > 0$, at epochs $T_k$, $k = 1, 2, ...$, that is, $N(t) = \max\{k, T_k \leq t\}$. The sequence $\{N(t), t \geq 0\}$, which is a non-decreasing function of $t$ and is a random process, is referred to as the counting process belonging to the point process $\{T_1, T_2, ..., T_r, T_1 \geq 0\}$. It is referred to as the simple counting process if one event can occur at a time. It is also referred to as the random point process if the time epochs $T_k$, $k = 1, 2, ...$, are random variables, with $P(\lim_{k\to\infty} T_k) = 1$. It is further referred to as the marked point process.

***Remark*** The increment $N(t+s) - N(t)$ will contain the number of events that occur in the interval $(t, t+s]$. That is why $\{N(t), t \geq 0\}$ is called the counting process, that is, counting the number of events that occurred by time $t$.

***Summarize***. We can summarize a counting process $\{N(t), t \geq 0\}$ as follows:

1. For each t, the values of $N(t)$ are nonnegative integers.
2. The function $N(t)$ is a non-decreasing function with respect to $t$.
3. The function $N(t)$ is a right continuous function.

The third property is merely a convention, and it interprets as follows: Let two events occur at times 5 and 8, that is, $N(5) = 1$, $N(8) = 1$ and

$$N(t) = \begin{cases} 1, & t \in (5, 8), \\ 0, & t < 5. \end{cases}$$

**Definition** Let us consider arrival epochs of tasks to a system. We denote the first $n$ inter-arrival times by $X_i$, $i = 1, 2, ..., n$, that is, $X_i - X_{i-1}$, $i = 1, 2, ..., n$, is the time elapsed between the $(i-1)$th arrival and the $i^{\text{th}}$, $i = 1, 2, \ldots, n$, arrival. We denote the $n$th arrival by $S_n$. Hence, for positive i.i.d. random variables, $X_1, X_2, \ldots, S_n = X_1 + X_2 + \ldots + X_n$. That is, each arrival epoch is the sum of $X_1, X_2, \ldots, X_n$. Then, not only is the sequence $\{S_n, n \geq 1\}$ a special case of a random walk, but also it is a sequence of epochs of reoccurrences, or recurrent or more commonly referred to as the renewal counting process, denoted by $\{N(t), n > 0\}$.

Because the process probabilistically starts over at each arrival epoch, the process is referred to as the renewal process. In this case, the process $\{S_n, n \geq 1\}$ should cross the threshold $T > 0$.

Note that the arrival epochs and the counting random variables are related as follows:

$$\{S_n \leq t\} = \{N(t) \geq n\} \text{ and } \{S_n > t\} = \{N(t) < n\}.$$

### 3.3.1 Poisson Process

**Definition** A counting process $N(t)$ is a Poisson process with parameter (or rate) $\lambda$ if the following conditions are true:

1. $N(0) = 0$.
2. It has independent increments, that is, if $(s_1, t_1) \cap (s_2, t_2) = \emptyset$, then the increment $N(t_1) - \mathrm{N}(\mathrm{s}_1)$ and $N(t_2) - \mathrm{N}(\mathrm{s}_2)$ are independent.
3. The number of events in any interval of length $t$ is the Poisson pmf with parameter $\lambda t$ that is,

$$P(N(t+s) - N(s) = j) = \frac{(\lambda t)^j}{j!}, \; j = 0, 1, 2, \ldots,$$

$$E[N(t+s) - N(s)] = \lambda t,$$

and

$$\lim_{\Delta \to 0} \{P(N(\Delta) = 1)\} = \lambda \Delta e^{-\lambda \Delta} \approx \lambda \Delta,$$

$$\lim_{\Delta \to 0} \{P(N(\Delta) \geq 2)\} = o(\Delta^2) \ll \lambda \Delta.$$

Last properties state that in an interval with a very small length, events occur singly with probability proportional to the length of the interval. This is why $\lambda$ is referred to as the rate.

***Example*** Consider a biased coin with high probability for a tail and low probability for a head to occur. We toss this coin $n$ times with $n$ to probability of success (head) as $P = \lambda/n < 1/2$, where $\lambda$ in a real number $< n$. Denote the number of heads in a period of time $[0, t]$ by $N_n(t)$. We leave it as an exercise to show that as $n$ approaches infinity, the number of heads occurred in $(s, t]$ is a binomial with number of trials equal to $n(t-s) \pm 2$ with $p = \lambda/n$, which converges to a Poisson pmf with parameter $t - s$.

***Example*** In this example, we show a property of a simple queue, which we will discuss later, as an application of a Poisson process. We consider arrival of tasks to a system to be a Poisson process with parameter λ We denote the inter-arrival times by $T_1, T_2, \ldots$, that is, $T_n$ is the time elapsed between the $(n-1)^{st}$ and $n^{\text{th}}$ arrivals, or the time between consecutive arrivals of tasks. We want to find the cumulative density function (CDF) of the inter-arrival times.

***Answer*** From the assumption for the arrivals to be Poisson, we conclude that pmf of arriving tasks follows a Poisson pmf with parameter $\lambda$. Hence, if the first arrival time is > *t*, there will not be any arrival task. In other words,

$$P(T_1 > t) = P(N(t) = 0) = e^{-\lambda t}.$$

Thus, $T_1$ is an exponential random variable with parameter $\lambda$. On the other hand,

$$\begin{aligned} \text{P}(T_2|T_1 = s) &= \text{P(No arrival in } (s, s+t]|T_1 = s) \\ &= \text{P}(N(t) = 0) = e^{-\lambda t}, \ s, t > 0. \end{aligned}$$

This is because the tasks in $(s, s+t]$ are not affected by what happens in $[0, s]$. Hence, $T_2$ is independent of $T_1$ and also is exponential with parameter $\lambda$. We leave it as an exercise by using mathematical induction, to show that $T_3$ is independent of $T_1$ and $T_2$ and so on.

**Theorem** *Let $T_1, T_2, \ldots$ be i.i.d. exponential random variables with parameter $\lambda$. We also let $S_n$ denote the sum of the first n of these random variables, $T_1, T_2, \ldots$ that is, $S_n = T_1, T_2 + \ldots + T_n$, which presents the waiting time for the* n*th arrival. We further assume that $N(t)$ denotes the arrival making the n largest such that $S_n \le t$. Then, the following statements are true:*

1. $E[S_n] = \frac{n}{\lambda}$.
2. *The pdf of $S_n$ is $\Gamma(n, \lambda)$, the two-parameter gamma pdf, that is,*

$$fS_n = \frac{\lambda^n t^{n-1}}{(n-1)!} e^{-\lambda t}.$$

***Proof***

1. Since $T_1, T_2, \ldots, T_n$ are independent and the expected value of each $T_j, j = 1, 2, \ldots, n$, is $1/\lambda$ and $E[S_n] = n/\lambda$.
2. We have

$$P(S_n > t) = P(N(t) < n) = \sum_{j=0}^{n-1} \frac{(\lambda t)^j}{j!} e^{-\lambda t}.$$

Differentiating with respect to $t$, we obtain:

$$\begin{aligned}
-fS_n &= \sum_{j=0}^{n-1} \frac{1}{j!}\left[-(\lambda t)^j \lambda e^{-\lambda t} + j(\lambda t)^{j-1} \lambda e^{-\lambda t}\right] \\
&= \lambda e^{-\lambda t} \sum_{j=0}^{n-1} \left[-\frac{(\lambda t)^j}{j!} + \frac{j(\lambda t)^{j-1}}{(j-1)!}\right] \\
&= -\lambda \frac{(\lambda t)^{n-1}}{(n-1)!} e^{-\lambda t}.
\end{aligned}$$

This completes the proof.

***Example*** Suppose occurrences of events in a process are distributed according to a Poisson pmf with parameter λ.

I. Find the probability that the eighth event occurs 2 or more time units after the seventh event.
II. Find the expected time of occurrence of the eighth event.
III. Find the probability that the eighth event occurs after the time epoch 10.
IV. Find the probability that two events occur in the time interval [1, 3] and three events in [2, 4]. [2, 4].

***Answer***

I. The question we have to answer in this item is finding

P(The eight event to occur later than 10)

The answer is yet $e^{-\lambda t}$. This is because one can restart the Poisson process at any event.

II. From previous theorem, we have

$$E[S_8] = \frac{8}{\lambda}.$$

III. We need to find $P$(The eight event to occur later than 10). This is equivalent to finding:

$$P(S_8) = P(N(10)) = P(N(10) < 8).$$

We have two choices as how to find this probability. That is,

$$\text{a.}\quad P(S_8 > 10) = \int_{10}^{\infty} \frac{(\lambda t)^7}{7!} \lambda e^{-\lambda t} dt,$$

or

$$b.\quad P(N(10) < 8) = \sum_{j=0}^{7} \frac{(10\lambda)^j}{7!} \lambda e^{-10\lambda}.$$

Using (a) with the aid of two-line codes of MATLAB:
L = 5; %L = lambda.
fun = @(t) (L/(factorial(7))).*((L.*t).^7).*(exp(-L.*t)).
p = integral(fun,10,inf).
we obtain the values for different values of $\lambda$ as shown in Table 3.1.

IV. To find the probability that two events occur in the time interval $[1, 3]$ and three events in $[2, 4]$, symbolically, is $P$(Two events in [1,3] and three events in [2,4]). To find this probability, we use the conditional probability as follows:

$$\begin{aligned}
&P(\text{Two events in } [1,3] \text{ and three events in } [2,4]) \\
&= \sum_{j=0}^{2} P(\text{Two events in } [1,3] \text{ and three events in } [2,4] \mid j \text{ events } in\,[2,3]) \\
&\quad \times P(j \text{ events } in[2,3]) \\
&= \sum_{j=0}^{2} P(2-j \text{ events in } [1,2] \text{ and } 3-j \text{ events in } [2,4] \text{ and } 3-j \text{ events } in\,[3,4]) \\
&\quad \times P(j \text{ events } in\,[2,3]) \\
&= \sum_{j=0}^{2} \frac{\lambda^{2-j}}{(2-j)!} e^{-\lambda} \frac{\lambda^{3-j}}{(3-j)!} e^{-\lambda} \frac{\lambda^j}{j!} e^{-\lambda} \\
&= \frac{\lambda^3(\lambda^2 + 6\lambda + 6)}{12} e^{-3\lambda}.
\end{aligned}$$

Again, for different values of $\lambda$, we can find the values of this probability in Table 3.2.

**Table 3.1** Occurrences of events in a waiting time for the 8th arrival with different parameters of $\lambda$

| $\lambda$ | 0.5 | 1 | 3 | 5 |
|---|---|---|---|---|
| $p(s_8 > 10)$ | 0.8666 | 0.2202 | 5.2337e−07 | 3.4640e−14 |

**Table 3.2** The probability of two events in [1, 3] and three events in [2, 4] with different $\lambda$

| $\lambda$ | 0.5 | 1 | 3 | 5 |
|---|---|---|---|---|
| P(2 events in [1, 3] and 3 events in [2, 4]) | 0.0215 | 0.0539 | 0.0092 | 1.9438e−04 |

## 3.4 Martingale Processes

In probability theory, martingale is a sequence of random variables (random process) in which in each part of time, the average value of the random variable for the next time in the sequence is equal to the value of the current random variable, provided that all previous values of the random variable are clear.

If $X$ is a sequence of random variables, then for any arbitrary time such as $n$ the following two conditions are hold:

$$E|X_n| < \infty$$
$$E[X_{n+1}|X_1, X_2, ..., X_n] = X_n.$$

It is clear that the mean of a random variable at time $n + 1$ is equal to the value at present $X_1, X_2, ..., X_n$ knowing its value in the previous steps $n$.

The increment of a random process $X = (X_t : t \in T)$ over an interval $[a, b]$ is the random variable $X_b - X_a$.

A random process is said to have independent increments if for any positive integer $n$ and any $t_0 < t_1 < ... < t_n$ in $T$, the increments $X_{t_1} - X_{t_0}, ..., X_{t_n} - X_{t_{n-1}}$ are mutually independent.

A random process $X = (X_t : t \in T)$ is called a martingale if $E[X_t]$ is finite for all $t$ and for any positive integer $n$ and $t_0 < t_1 < ... < t_n < t_{n+1}$,

$$E[X_{t_{n+1}} X_{t_1}, ..., X_{t_n}] = X_{t_n}$$

or, equivalently,

$$E[X_{t_{n+1}} - X_{t_n}|X_{t_1}, ..., X_{t_n}] = 0.$$

If $t_n$ is interpreted as the present time, then $t_{n+1}$ is a future time and the value of $(X_{t_1}, ..., X_{t_n})$ represents information about the past and present values of $X$. With this interpretation, the martingale property is that the future increments of $X$ have conditional mean zero, given the past and present values of the process.

An example of a martingale is the following. Suppose a gambler has initial wealth $X_0$. Suppose the gambler makes bets with various odds, such that, as far as the past history of $X$ can determine, the bets made are all for fair games in which the expected net gains are zero. Then if $X_t$ denotes the wealth of the gambler at any time $t_0$, then $(X_t : t \geq 0)$ is a martingale.

Suppose $(X_t)$ is an independent increment process with index set $T = \mathbb{R}^+$ or $T = \mathbb{Z}^+$, with $X_0$ equal to a constant and with mean zero increments. Then $X$ is a martingale, as we now show. Let $t_0 < t_1 < ... < t_{n+1}$, be in $T$. Then $(X_{t_1}, ..., X_{t_n})$ is a function of the increments $X_{t_1} - X_{t_0}, ..., X_{t_n} - X_{t_{n-1}}$ and hence it is independent of the increment $(X_t : t \geq 0)$. Thus

$$E[X_{t_{n+1}} - X_{t_n}|X_{t_1}, ..., X_{t_n}] = E[X_{t_{n+1}} - X_{t_n}] = 0.$$

The random walk $(X_n : n \geq 0)$ arising in the gambler's ruin problem is an independent increment process, and if p/12 is also a martingale.

## 3.5 Brownian Motion

A process $W = (W_t : t \geq 0)$ is called Brownian motion process if the following conditions hold:

i. $P(W_0 = 0) = 1$.
ii. The increments *of* $W$ are independent.
iii. $W_t - W_s \sim N(0, t - s)$ for $t \geq s$.
iv. $P(W_t$ is a continuous function of $t) = 1$ (Fig. 3.3).

The mean, correlation, and covariance functions of a Brownian motion $W$ are given by

$$\mu_W(t) = E[W_t] = E[W_t - W_0] = 0$$

and, for $s \leq t$,

$$\begin{aligned} R_W(s,t) &= E[W_s W_t] \\ &= E[(W_t - W_0)(W_s - W_0 + W_t - W_s)] \\ &= E[(W_t - W_0)^2] = s \end{aligned}$$

so that, in general,

$$C_W(s,t) = R_W(s,t) = \min(s,t).$$

A Brownian motion is Gaussian, because if $0 = t_0 \leq t_1 \leq \dots \leq t_n$, then each coordinate of the vector $(W_{t_1}, \dots, W_{t_n})$ is a linear combination of the $n$ independent Gaussian random variables $(W_{t_i}, \dots, W_{t_{i-1}} : 1 \leq i \leq n)$. Thus, properties (I-III) imply that $W$ is a Gaussian random process with $\mu_W = 0$ and $R_W(s,t) = \min(s,t)$. In fact,

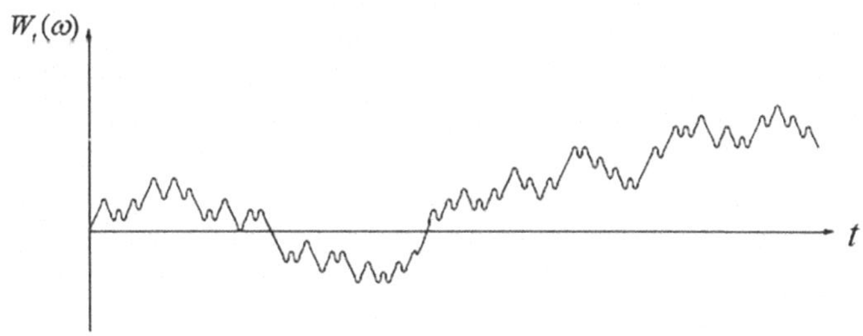


**Fig. 3.3** A sample path of Brownian motion

the converse is also true. If $W = (W_t : t \geq 0)$ is a Gaussian random process with mean zero and $R_W(s, t) = \min(s, t)$, then (I-III) are true.

Property (IV) does not come automatically. For example, if $W$ is a Brownian motion and if $U$ has a uniform distribution; Unif(0, 1) independent of $W$, let $\tilde{W}$ be defined by

$$\tilde{W}_t = W_t + I_{\{U=t\}}.$$

## 3.6 Stationarity

Consider a random process $X = (X_t : t \in T)$ such that either $T = \mathbb{Z}$ or $T = \mathbb{R}$. Then $X$ is said to be stationary if for any $t_1, ..., t_n$ and $s$ in $T$, the random vectors $(X_{t_1}, ..., X_{t_n})$ and $(X_{t_1+s}, ..., X_{t_n+s})$ have the same distribution.

Suppose $X$ is a stationary second order random process. (Recall that second order means that $E[X_t^2] < \infty$ for all $t$.) Then by the $n = 1$ part of the definition of stationarity, $X_t$ has the same distribution for all $t$. In particular, $\mu_X(t)$ and $E[X_t^2]$ do not depend on $t$. Moreover, by the $n = 2$ part of the definition $E[X_{t_1}X_{t_2}] = E[X_{t_1+s}X_{t_2+s}]$ for any $s \in T$. If $E[X_t^2] < \infty$ for all $t$, then $E[X_{t+s}]$ and $R_X(t_1 + s, t_2 + s)$ are finite and both do not depend on $s$.

A second order random process $(X_t : t \in T)$ with $T = \mathbb{Z}$ or $T = \mathbb{R}$. is called wide sense stationary (WSS) if

$$\mu_X(t) = \mu_X(s + t) \text{ and } R_X(t_1, t_2) = R_X(t_1 + s, t_2 + s)$$

for all $t, s, t_1, t_2 \in T$. As shown above, a stationary second order random process is WSS. Wide sense stationarity means that $\mu_X$ is a finite number, not depending on $t$, and $R_X(t_1, t_2)$ depends on $t_1, t_2$ only through the difference $t_1 - t_2$. By a convenient and widely accepted abuse of notation, if $X$ is WSS, we use $\mu_X$ to be the constant and $R_X$ to be the function of one real variable such that

$$\begin{aligned} E[X_t] &= \mu_X \; t \in T, \\ E[X_{t_1}X_{t_2}] &= R_X(t_1 - t_2) \; t_1, t_1 \in T. \end{aligned}$$

The dual use of the notation $R_X$ if $X$ is WSS leads to the identity $R_X(t_1, t_2) = R_X(t_1 - t_2)$. As a practical matter, this means replacing a comma by a minus sign. Since one interpretation of $R_X$ requires it to have two arguments, and the other interpretation requires only one argument, the interpretation is clear from the number of arguments. Some brave authors even skip mentioning that $X$ is WSS when they write: "Suppose $(X_t : t \in \mathbb{R})$ has mean $\mu_X$ and correlation function $R_X(\tau)$, " because it is implicit in this statement that $X$ is WSS.

Since the covariance function $C_X$ of a random process $X$ satisfies

$$C_X(t_1, t_2) = R_X(t_1, t_2) - \mu_X(t_1)\mu_X(t_2),$$

if $X$ is WSS then $C_X(t_1, t_2)$ is a function of $t_1 - t_2$. The notation $C_X$ is also used to denote the function of one variable such that $C_X(t_1 - t_2) = Cov(X_{t_1}, X_{t_2})$. Therefore, if $X$ is WSS then $C_X(t_1 - t_2) = C_X(t_1, t_2)$. Also, $C_X(\tau) = R_X(\tau) - \mu_X^2$, where in this equation $\tau$ should be thought of as the difference of two times, $t_1 - t_2$.

In general, there is much more to know about a random vector or a random process than the first and second moments. Therefore, one can mathematically define WSS processes that are spectacularly different in appearance from any stationary random process. For example, any random process $(X_k : k \in \mathbb{Z})$ such that the $X_k$ are independent with $E[X_k] = 0$ and $Var[X_k] = 1$ for all $k$ is WSS. To be specific, we could take the $X_k$ to be independent, with $X_k$ being $N(0, 1)$ for $k \leq 0$ and with $X_k$ having pmf

$$p_{X,1}(x, k) = P(X_k = x) = \begin{cases} \frac{1}{2k^2} \, x \in \{k, -k\} \\ 1 - \frac{1}{k^2} \text{ if } x = 0 \\ 0 \text{ else} \end{cases}$$

for $k \geq 1$ (Fig. 3.4).

***Example*** Consider the random process $\{X(t),\ t \in \mathbb{R}\}$ defined as

$$X(t) = \cos(t + u),$$

where $U \sim uniform(0, 2\pi)$. Then $X(t)$ is a WSS process.

We need to check two conditions:

$$\text{(i)} \quad \mu_X(t) = \mu_X, \ \text{ for all } t \in \mathbb{R}, \ \text{ and}$$

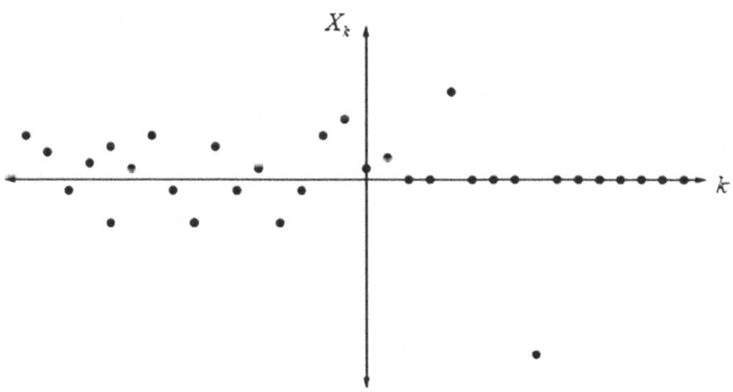


**Fig. 3.4** A typical sample path of WSS processes

$$\text{(ii)}\quad R_X(t_1, t_2) = R_X(t_1 - t_2), \text{ for all } t_1, t_2 \in \mathbb{R}.$$

We have

$$\begin{aligned}
\mu_X(t) &= E[X(t)] \\
&= E[\cos(t + U)] \\
&= \int_0^{2\pi} \cos(t + u)\frac{1}{2\pi}du \\
&= 0, \quad \text{for all } t \in \mathbb{R}.
\end{aligned}$$

We can also find $R_X(t_1, t_2)$ as follows:

$$\begin{aligned}
R_X(t_1, t_2) &= E[X(t_1)X(t_2)] \\
&= E[\cos(t_1 + U)\cos(t_2 + U)] \\
&= E[\frac{1}{2}\cos(t_1 + t_2 + 2U) + \frac{1}{2}\cos(t_1 - t_2)] \\
&= E[\frac{1}{2}\cos(t_1 + t_2 + 2U)] + E[\frac{1}{2}\cos(t_1 - t_2)] \\
&= \int_0^{2\pi} \cos(t_1 + t_2 + u)\frac{1}{2\pi}du + \frac{1}{2}\cos(t_1 - t_2) \\
&= 0 + \frac{1}{2}\cos(t_1 - t_2) \\
&= \frac{1}{2}\cos(t_1 - t_2), \quad \text{for all } t_1, t_2 \in \mathbb{R}.
\end{aligned}$$

As we see, both conditions are satisfied, thus $X(t)$ is a WSS process.

## 3.7 Exercises

(1) Let $X(t)$ be a zero-mean WSS Gaussian process with $R_X(\tau) = e^{-\tau^2}$, *for all* $\tau \in \mathbb{R}$.

- Find $P(X(1) < 1)$.
- Find $P(X(1) + X(2) < 1)$.

(2) Let $X(t)$ and $Y(t)$ be two jointly WSS random processes. Consider the random process $Z(t)$ defined as

$$Z(t) = X(t) + Y(t).$$

Show that $Z(t)$ is WSS.

(3) Consider the random process $\{X_n, n = 0, 1, 2, ...\}$ in which $X_i's$ are i.i.d. standard normal random variables.

- Find $f_{X_n}(x)$, for $n = 0, 1, 2, ...$
- Find $f_{X_n X_m}(x_1, x_2)$, for $n \neq m$.

(4) You have 1000 dollars to put in an account with interest rate $R$, compounded annually. That is, if $X_n$ is the value of the account at year $n$, then

$$X_n = 1000(1 + R)^n, \text{ for } n = 0, 1, 2, ...$$

- Find all possible sample functions for the random process $\{X_n, n = 0, 1, 2, ...\}$
- Find the expected value of your account at year three. That is, find $E[X_3]$.

(5) Let $\{X(t), t \in [0, \infty)\}$ be defined as

$$X(t) = A + Bt, \text{ for all } t \in [0, \infty),$$

where $A$ and $B$ are independent normal $N(1, 1)$ random variables.

- Find all possible sample functions for this random process.
- Define the random variable $Y = X(1)$. Find the PDF of $Y$.
- Let also $Z = X(2)$. Find $E[Y]$.

(6) Find the mean functions for the random processes given in Exercises 4 and 5.
(7) Find the correlation functions and covariance functions for the random processes given in Exercises 4 and 5.
(8) Let $A$, $B$ and $C$ be independent normal $N(1, 1)$ random variables. Let $\{X(t), t \in [0, \infty)\}$ be defined as

$$X(t) = A + Bt, \text{ for all } t \in [0, \infty).$$

Also, let $\{Y(t), t \in [0, \infty)\}$ be defined as

$$Y(t) = A + Ct, \text{ for all } t \in [0, \infty).$$

Find $R_{XY}(t_1, t_2)$ and $C_{XY}(t_1, t_2)$ for $t_1, t_2 \in [0, \infty)$.
$R_{XY}(t_1, t_2)$ and $C_{XY}(t_1, t_2)$ for $t_1, t_2 \in [0, \infty)$.

(9) Consider the discrete-time random process $\{X(n), n \in \mathbb{Z}\}$ in which the $X(n)'s$ are i.i.d. with CDF $F_{X(n)}(x) = F(x)$. Show that this is a (strict-sense) stationary process.

(10) If $X(t)$ is a Poisson process with intensity $\lambda$, then for all $t > s \geq 0$, we have

$$X(t) - X(s) \sim Poisson(\lambda(t - s)).$$

Show that $X(t)$ is mean-square continuous at any time $t \geq 0$. $t \geq 0$.

(11) Consider a random process $X(t)$ and its derivative, $X'(t) = \frac{d}{dt}X(t)$. Assuming that the derivatives are well-defined, show that

$$R_{XX'}(t_1, t_2) = \frac{\partial}{\partial t_2} R_X(t_1, t_2).$$

(12) Consider the Gaussian random processes $\{X(t), t \in \mathbb{R}\}$. If $X(t)$ is WSS, then $X(t)$ is a stationary process.

(13) Let $Y_1, Y_2, Y_3, ...$ be a sequence of i.i.d. random variables with mean $E[Y_i] = 0$ and $Var[Y_i] = 4$. Define the discrete-time random process $\{X(n), n \in \mathbb{N}\}$ as

$$X(n) = Y_1 + Y_1 + ... + Y_1 \text{ for all } n \in \mathbb{N}.$$

Find $\mu_X(n)$ and $R_X(m, n)$, for all $n, m \in \mathbb{N}$.$\mu_X(n)$ and $R_X(m, n)$, for all $n, m \in \mathbb{N}$.

(14) For any $k \in \mathbb{Z}$, define the function $g_k(t)$ as

$$g_k(t) = \begin{cases} 1 0 < t \leq k + 1, \\ 0 \text{ otherwise.} \end{cases}$$

Now, consider the continuous-time random process $\{X(t), t \in \mathbb{R}\}$ defined as

$$X(t) = \sum_{k=-\infty}^{+\infty} A_k g_k(t),$$

where $A_1, A_2, ...$ are i.i.d. random variables with $E[A_k] = 1$ and $Var[A_k] = 1$. Find $\mu_X(t)$, $R_X(s, t)$, and $C_X(s, t)$ for all $s, t \in \mathbb{R}$.

(15) Let $X(t)$ be a continuous-time WSS process with mean $\mu_X = 1$ and

$$R_X(\tau) = \begin{cases} 3 - |\tau| \ -2 \leq \tau \leq 2, \\ 1 \text{ otherwise.} \end{cases}$$

- Find the expected power in $X(t)$.
- Find $E\big[(X(1) + X(2) + X(3))^2\big]$.

(16) $X(t)$ be a continuous-time WSS process with mean $\mu_X = 0$ and

where $\delta(\tau)$ is the Dirac delta function. We define the random process $Y(t)$ as

$$Y(t) = \int_{t-2}^{t} X(u)du.$$

- Find $\mu_Y(t) = E[Y(t)]$.
- Find $R_{XY}(t_1, t_2)$.

(17) Let $X(t)$ be a Gaussian process with $\mu_Y(t) = t$,and $R_X(t_1, t_2) = 1 + 2t_1t_2$, for all $t, t_1, t_2 \in \mathbb{R}$. Find $P(2X(1) + X(2) < 3)$.

## Reference

1. Tulcea AI (1968) Ergodic theory and information (Patrick Billingsley). SIAM Rev 10(4):459

## *Uncited References*

2. Scott M (2013) Applied stochastic processes in science and engineering, University of Waterloo, eBook
3. Gallager RG (2013) Stochastic processes: theory for applications. Cambridge University Press
4. Wong E, Hajek B (2012) Stochastic processes in engineering systems. Springer Science & Business Media
5. Rosen KH (2007) Discrete mathematics and its applications. The McGraw Hill Companies
6. Graham RL, Knuth DE, Patashnik O (1994) Concrete math: a foundation for computer science
7. Tucker A (1994) Applied combinatorics. Wiley
8. Peebles Jr PZ (1987) Probability random variables, and random signal principles
9. Gubner JA (2006) Probability and random processes for electrical and computer engineers. Cambridge University Press
10. Kobayashi H, Mark BL, Turin W (2011) Probability, random processes, and statistical analysis: applications to communications, signal processing, queueing theory and mathematical finance. Cambridge University Press
11. Jones PW, Smith P (2017) Stochastic processes: an introduction. Chapman and Hall/CRC Press
12. Stirzaker D (2005) Stochastic processes and models. OUP Oxford
13. Buckley JJ (2005) Fuzzy probabilities: new approach and applications, vol 115. Springer Science & Business Media
14. Ross TJ, Booker JM, Parkinson WJ (eds) (2002) Fuzzy logic and probability applications: bridging the gap. Society for Industrial and Applied Mathematics
15. Klir GJ, St. Clair U, Yuan B (1997) Fuzzy set theory: foundations and applications. Prentice-Hall, Inc.
16. Harchol-Balter M (2013) Performance modeling and design of computer systems: queueing theory in action. Cambridge University Press
17. Borovkov A (2012) Stochastic processes in queueing theory, vol 4. Springer Science & Business Media
18. Kleinrock L (1975) Queueing systems: theory. Wiley

19. Feller W (1991) An introduction to probability theory and its applications, vol 2, vol 81. Wiley
20. Shafer G, Vovk V (2005) Probability and finance: it's only a game! vol 491. Wiley
21. Rachev ST, Hoechstoetter M, Fabozzi FJ, Focardi SM (2010) Probability and statistics for finance, vol 176. Wiley

The manufacturer's authorised representative in the EU is Springer Nature Customer Service Centre GmbH, Europaplatz 3, 69115 Heidelberg, Germany. If you have any concerns regarding our products, please contact ProductSafety@springernature.com

Printed and bound by CPI Group (UK) Ltd, Croydon, CR0 4YY
31/07/2026
02182962-0001